Python for Linguists

Specifically designed for linguists, this book provides an introduction to programming using Python for those with little to no experience with coding. Python is one of the most popular and widely used programming languages as it's also available for free and runs on any operating system. All examples in the text involve language data and can be adapted or used directly for language research. The text focuses on key language-related issues: searching, text manipulation, text encoding and internet data, providing an excellent resource for language research. More experienced users of Python will also benefit from the advanced chapters on graphical user interfaces and functional programming.

MICHAEL HAMMOND is Professor of Linguistics and Human Language Technology at the University of Arizona. His previous titles include *Programming for Linguists: Perl for Language Professionals* (2003) and *Programming for Linguists: Java Technology for Language Professionals* (2002).

Python for Linguists

Michael Hammond
University of Arizona

CAMBRIDGE
UNIVERSITY PRESS

CAMBRIDGE
UNIVERSITY PRESS

University Printing House, Cambridge CB2 8BS, United Kingdom

One Liberty Plaza, 20th Floor, New York, NY 10006, USA

477 Williamstown Road, Port Melbourne, VIC 3207, Australia

314–321, 3rd Floor, Plot 3, Splendor Forum, Jasola District Centre,
New Delhi – 110025, India

79 Anson Road, #06–04/06, Singapore 079906

Cambridge University Press is part of the University of Cambridge.

It furthers the University's mission by disseminating knowledge in the pursuit of education, learning, and research at the highest international levels of excellence.

www.cambridge.org
Information on this title: www.cambridge.org/9781108493444
DOI: 10.1017/9781108642408

First published 2020

A catalogue record for this publication is available from the British Library.

ISBN 978-1-108-49344-4 Hardback
ISBN 978-1-108-73707-4 Paperback

Contents

Preface

This is a book on how to program for linguistic purposes using the Python programming language. In this preface we outline our goals, justify using Python to achieve them, and explain how best to use this book.

Why Do Linguists Need to Learn How to Program?

Programming is an extremely useful skill in many areas of linguistics and in other language-related fields like speech and hearing sciences, psychology, psycholinguistics, and quantitative literary studies.

Within linguistics, it used to be the case that programming skills were required only for computational linguists, but this is far from true these days. Programming now is used in phonology, syntax, morphology, semantics, pragmatics, psycholinguistics, phonetics, discourse analysis, essentially every area of linguistic investigation. This change reflects broader methodological changes in the field, a response to the fact that (i) more and more data are available electronically, and (ii) we have much richer techniques for examining and manipulating massive amounts of electronic data.

Here are some examples of what you can do with fairly modest programming skills:

- Build a simple list of occurring words from a text file written in some language along with the frequency of those words.
- Find items for psycholinguistic or phonetic experiments from text resources, e.g., frequent two-syllable words that begin with a three-consonant cluster and don't otherwise contain nasal consonants.
- Construct every possible one-syllable word given a set of possible onsets, vowels, and codas.
- Construct every possible two-word compound given a list of words.
- Find the average number of words per sentence from a text corpus and find the longest sentence in that corpus.

- Build a model of syllabification for some language and syllabify candidate words.
- Find the average length and amplitude for some set of sound files.

This is just a small sample of the sorts of things that programming might help you do as part of your work with language.

Why Is Python a Good Choice?

There is a wide variety of programming languages you might choose to learn and work with. Every language has its virtues, the things that make it a good choice for this or that purpose. The choice of language is a function of three basic factors.

First, what is it that you want to do? Some languages are optimized for different sorts of goals, e.g., writing applications, developing system tools, and scripting. Second, what kind of programming experience do you have? Some languages are relatively easy to pick up and others, not so much. Finally, what kind of programming style do you want to use? Different languages lend themselves to different kinds of programming approaches, e.g., procedural, functional, object-oriented, and parallel.

In this book, we assume that you want to write programs that will let you answer questions about language, programs that may be run only by you. Thus programming languages like `java`, objective `c`, or `c#` that allow you to write full applications are not optimal choices. In addition, we assume that you have little or no prior programming experience. Thus interesting and challenging programming systems like `Haskell`, `Lisp`, or `Prolog` are best left for later.

Consequently, we will use the Python programming language. There are specific reasons for this choice:

- Python is *extremely* widely used, so you most likely have friends, colleagues, or classmates who use it and so can help you if needed. There are myriad resources on the web that can help as well.
- Python has stable and clear semantics. This means that the meaning and use of Python elements is clearly defined, and you can rely on your programs working as you intend. Similarly, program examples in this text should work exactly as shown on your system.
- Modules. There are tons of optional modules that others have written with useful functions and objects to simplify your programming tasks.
- Python is practical. Python is widely used in many areas, so your language-related programming skills may help you in other domains.

- NLTK. The Natural Language Toolkit is a freely available suite of modules tailored for working with language. You can use these for high-level statistical natural language processing or for simple language-related tasks.

There are some challenges to working with Python too. The biggest is that Python is an object-oriented (OO) programming language, and using it requires that you at least understand what objects are. The language thus lends itself to an OO programming style, but you need not use that specific style at first.

Our approach in this book is to first ignore OO aspects of the language. As we proceed, we introduce what you need to know of the OO system to make use of the concepts introduced to that point. Finally, in the latter part of the book, we explain OO programming in depth. Ultimately, we leave it up to you to decide how much OO programming is necessary for your programming goals.

How This Book Is Different

There are any number of book-length introductions to Python out there. How is this one different?

First, this book is written for linguists and other people who work with language data. What this means is that we give you examples that should make sense to you. If you have specific programs you want to write right away, you may even find some snippet here that (almost) does what you want and can be easily adapted to your purposes.

We continue with the language focus throughout the book. This means that, rather than trying to learn some programming concept exemplified in a program that has no relation to your goals, you can learn critical concepts with programs that are comprehensible and, we hope, useful as well.

Most chapters conclude with an extended example that shows how a larger program should be developed, using the concepts learned in the chapter, and with a focus on some language-related task. All chapters conclude with exercises, and all the exercises are linguistic in their orientation as well.

Another consequence of this approach is that we take a linguistic approach to the structure of Python. As much as possible, we treat it as a language with a syntax and a semantics.

Overview of the Book

The structure of this book is roughly as follows.

First, we introduce the basic syntax and semantics of the language: the primitive elements of the language and how those elements can be combined to make legal statements and larger structures.

As we introduce those topics, we elaborate the imperative semantics of the language, how we can use the specific Python language components covered to achieve different programming goals.

We next consider specific language-related tasks in depth: searching text, manipulating text, internet data, and text encodings.

Finally, we conclude with more advanced discussions of Python objects and OO programming generally, GUI programming, and functional programming. There is a brief appendix outlining the NLTK system as well.

How to Use the Book

The most important thing about using this book is that you should run the programs as you proceed. You can either download them from the book website[1] or type them in yourself.

If you have the patience for it, it is much better to type the programs in. This will really help you to notice aspects of the code you might not otherwise see and make the coding process more familiar to you. This will be frustrating and you will make errors as you type things in, but figuring out these errors will really help you learn the material.

It's also extremely important that you understand the code. You should make sure you understand each code snippet that's given before you go on – how it works, why it does what it does.

To this end, another really useful thing to do as you proceed is to play with the code. Tweak it in different ways, either to do something you'd rather it do or just to see what happens. (As you'll see below, the only time you *don't* want to do this is when you're performing file input – output operations where you can accidentally damage or lose things on your computer.)

Acknowledgments

Thanks to Sam Johnston, Dan Jurafsky, Nick Kloehn, and Ben Martin for never letting me forget the virtues of Python.

Thanks to the students in my Linguistics 408/508 course for letting me try out this material with them. Thanks to Damian Romero Diaz for helpful comments.

Thanks to Diane Ohala and Joey Rousos-Hammond for their love and support throughout.

All errors are my own.

[1] `www.u.arizona.edu/~hammond/`

1 Interacting with Python and Basic Functions

In this chapter, we introduce the different ways you can interact with Python generally and some simple things you can do right from the get-go.

We start with how to install Python and issues concerned with different versions of the language. We then turn to how to invoke the interactive environment and the things you can do there. Finally, we briefly outline what it means to write a program.

1.1 Installing and Using Python

Python is a programming language. It is a way for you to talk to your computer, a way for you to get your computer to do things for you. It's basically a specialized language that is optimized for clarity for you and for converting into the internal language that your computer actually uses. You might think of it as a solution to a translation problem. You speak one language, and your computer speaks another. Python is a compromise language between the two. For you to program, you must translate your intentions into Python. For the computer to respond appropriately, it must translate Python into its own internal language.

To use Python it is not enough to simply translate your goals into Python. You must also install specialized software on your computer that will translate your Python code into what the computer can work with. We will refer to this software as a *Python installation.*

Sometimes this software is automatically part of your operating system. For example, Python is part of every MacOS and Linux installation. Sometimes this software has to be installed by you. Sometimes this software is free; some companies charge for it. My recommendation is that you at least start with a free version. If later on, when you really know what you're doing with Python, some proprietary version offers features that you feel are necessary, then that's the time a purchase may be warranted.

A related issue here is that there are different versions of Python. The language was first released in 1991, and there have been a number of different versions since then offering changes and improvement. As I write this, the current version is 3.6.2.

When you install Python, you may have a choice about versions. My recommendation is that you install something in the version 3 family. There are significant differences between versions 2 and 3, and version 3 is what we use here.

In general, you want the most recent version possible, but there are sometimes reasons not to do that. Some third-party modules are not available for all versions, and so if there is a particular module that is important to you, you may want to make sure you have a version of Python that works with it.

There are a number of free versions of Python that are available for Mac and Windows. For example, one widely used version is Anaconda.[1] Another widely used system is ActiveState.[2] Python.org[3] maintains a list of Python distributions and distribute their own as well. I've already mentioned that Python is part of any MacOS, but if you need a specific version of Python for your purposes, you can get those from MacPorts[4] or Homebrew.[5]

1.2 The Interactive Environment

There are at least four ways you can invoke Python:

(i) **interactive environment** terminal or Python window
(ii) **Idle** Python integrated development environment
(iii) **edit and run** write a program that you run in the terminal window
(iv) **Jupyter notebooks** write code that runs interactively in a web browser

Ultimately, we will want to write programs that we then run in the terminal window, but at this stage we will confine our attention to the interactive environment. This will allow us to play with Python so we can understand the basics before we move on to writing programs.

If Python is on your system – or you've properly installed it – you can start the interactive environment by simply typing `python` in the terminal window on a Mac or choosing Python from the start menu on Windows. This will produce a response like this:

```
> python
Python 3.4.6 ...
Type "help", "copyright", "credits" or
"license" for more information.
>>>
```

[1] www.continuum.io
[2] www.activestate.com
[3] wiki.python.org/moin/PythonDistributions
[4] www.macports.org
[5] brew.sh

The version of Python that you're using is displayed along with additional system information. All of this is followed by the prompt >>>. Your commands are typed at that prompt.

Before going any further, let's set out the most important commands that work here:

Quit: `quit()`. Type this at the >>> prompt to exit the interactive environment. Typing `^d` (`control-d`) has the same effect.

Help: `help()`. Typing this enters the online help system where you can get help on many aspects of using Python. While this is extremely helpful, the responses it provides may not be terribly useful at this stage. You exit the help system and return to the interactive environment with the `return` key.

Interrupt: `^c` (`control-c`). When we start playing with the system, you will occasionally get stuck in the middle of a command or while some command is running. If you do get stuck, or if some command seems to be running forever, you can often regain control and return to the >>> prompt by typing `^c`.

1.3 Basic Interactions

We will generally interact with Python by writing programs and then running them from the command line. At this point, however, let's try to understand Python a bit better in the interactive environment. In this environment, you can type legal Python statements and they are immediately evaluated. For example, you can perform mathematical calculations like multiplication, addition, and exponentiation:

```
>>> 4 * 7
28
>>> 3 + 2.9
5.9
>>> 9 ** -3
0.0013717421124828531
>>> 7 - 15
-8
```

As linguists, we will want to operate on words and sentences, and these must be entered in single or double quotes. In general terms, we will refer to these as *character strings*.

```
'This is a sentence'
"This is another one"
```

Interestingly, some of the mathematical operators above can be used with character strings as well and have different effects. With numbers + is addition, but with strings it concatenates. With numbers * is multiplication, but with a string and a number it repeats the string:

```
>>> 'phon' + 'ology'
'phonology'
>>> 'phon' * 6
'phonphonphonphonphonphon'
```

There are some functions that operate directly on strings. The `len()` function returns the number of characters in a string.

```
>>> len('phonetics')
9
```

The `type()` function can operate on strings or numbers and returns the general "type" of the object it applies to.

```
>>> type(3)
<class 'int'>
>>> type(7.2)
<class 'float'>
>>> type('phoneme')
<class 'str'>
```

You can see that Python distinguishes integer numbers from floating point numbers (numbers with a decimal point) and from character strings.

Note now that the quotes surrounding a character string are essential and distinguish strings from other types. If you leave them out, Python will typically give you an error:

```
>>> phoneme
  Traceback (most recent call last):
  File "<stdin>", line 1, in <module>
  NameError: name 'phoneme' is not defined
```

If you put them in with numbers, then you change the basic type of the object. Consider the difference here:

```
>>> 7 + 4
11
>>> '7' + '4'
'74'
```

Recall that the + operator can apply to strings or numbers, but has different effects with each. When it applies to numbers, it performs addition. When it applies to strings, it performs string concatenation. We see here that when numbers are put in quotes they are treated as strings rather than as numbers.

There are functions for converting back and forth between character strings, integers, and floating point numbers:

```
>>> int('7')
7
>>> float('3.6')
3.6
>>> str(17)
'17'
```

Operators and functions can be combined as well. For example:

```
>>> (7 + 4) * 2
22
>>> str(7) + str(4)
'74'
>>> int(str(7)) ** 3
343
```

The format we've used for functions is to invoke the function by putting parentheses after its name, e.g., `str()`. There is another kind of function that we will frequently need that has a different syntax. These are *methods*, functions that are very specifically associated with some particular data type. We'll have a lot more to say about these in Chapter 10, but for now we just want to alert you that they exist and show you their syntax. To invoke one of these, the method name and parentheses are suffixed to the relevant object with a period. For example, the string method `upper()` returns an uppercase version of a string. We invoke it as follows:

```
>>> 'this is not a pipe'.upper()
'THIS IS NOT A PIPE'
```

There are methods that take additional arguments as well. For example, the method `count()` returns the number of instances of its argument in the string.

```
>>> 'this is not a pipe'.count('i')
3
```

Methods are *not* the same thing as functions; the different syntax above is required. Invoking them as if they were normal functions results in errors:

```
>>> upper('this is not a pipe')
Traceback (most recent call last):
  File "<stdin>", line 1, in <module>
NameError: name 'upper' is not defined

>>> count('this is not a pipe','i')
Traceback (most recent call last):
  File "<stdin>", line 1, in <module>
NameError: name 'count' is not defined
```

1.4 Edit and Run

In general, one uses Python to write programs to be run from the command line. For example, we can write a silly little program that concatenates two strings and prints the output. We would edit a text file, say with the name `interact1.py`, to contain the following:

```
print(4+7)
print('that should be 11')
```

interact1.py

Some parts of this are familiar from what we've covered so far, but some are not. We've added the `print()` function here, which simply prints its argument. Don't worry yet about the details; at this point, we just want several lines of code for our program.

It's important that you create this file with a simple text editor, not a full-on word processor like Microsoft Word. For writing code, I use Vim on a Mac, but Emacs (or Aquamacs) is a simple and free alternative. For Windows, there are various free ports of Emacs as well, but a number of other alternatives exist. It's most useful if the editor you use does *syntax colorizing*. This highlights different terms and strings in your Python code and makes it easier to spot errors. There are all sorts of other bells and whistles that different text editors offer, but the only essential ones at this point are (i) that the program be able to edit simple text files, and (ii) that it does syntax colorizing for Python.[6]

Once you've created the program with whatever text editor you opt for, there are a number of ways to invoke it. The simplest is to

[6] I strongly encourage you *not to* spend money on a text editor, at least at this stage. There are a huge number of free options.

type `python interact1.py` in the terminal or DOS window as follows:

```
> python interact1.py
11
That should be 11
```

If you invoke the program as above, it's important that the program be in the same directory that the terminal window is in. To find out what directory you are in, you can type `pwd` at the terminal prompt. To see if this program file is in that directory, you can use the `ls` command at the terminal prompt (`dir` for Windows). If it is not, you can either move it there or you can instruct the terminal to switch to another directory with the `cd` command (followed by the path to the directory you want to be in).

One can also add comments to a code file. Comments are lines of code that remind the programmer of what the code does or should do. Comments are marked with a # on their left. Everything on the same line after that # has no effect on the program. For example, we might tweak our program above like this:

```
print(4+7) #this does some math

#the following prints a string
print('that should be 11')
```

interact2.py

Notice that comments can occur on their own line or on the right side of an actual line of code.

1.5 Summary

In this chapter we have discussed how to install Python and the different versions that exist. We've also talked about how to run Python code and exemplified some basic commands in the interactive environment. We also briefly showed how to write and run a program.

1.6 Exercises

1.1 Write commands that print out your first name, the number of characters in that name, your last name, the number of characters in that name, and then concatenate and print the two names (with a space).

1.2 Write a line of code that will calculate how many times the string 'at' occurs in this sentence.

1.3 What's the difference between these?

```
'This is Mike'.upper().lower()
'This is Mike'.lower().upper()
```

1.4 Why does `upper('This is a cat')` not work?

1.5 What does `help(help)` do in the interactive environment?

1.6 We used the mathematical operator `**` on page 3 above without explanation. Play around with it and say what it does.

1.7 In math, `6 + 2` and `2 + 6` *mean* the same thing. We've seen that + and `*` can be used with strings too. What happens if arguments are reversed when strings are involved? Do those expressions *mean* the same thing?

1.8 Use the `help()` function to find out about the `round()` command. Explain and demonstrate how to use it with more than one argument to produce different results.

1.9 For each of the following, explain whether it's true or false and why:

(a) `'hat' == "hat"`
(b) `hat == 'hat'`
(c) `1/3 == .33`
(d) `'three' > 'two'`
(e) `2 + 2 = 4`

1.10 **Web**: Snoop around on the web and figure out what Jupyter notebooks are and how they work. Create a Jupyter notebook that answers one of the questions above.

2 Data Types and Variables

In this chapter, we look more closely at Python data types and variables. We'll discuss the following:

- Numbers: integers and floats
- Strings
- Lists
- Tuples
- Dictionaries

This discussion cannot proceed without considering the notion of *mutability*. Basically, some kinds of data cannot be changed once they have been created, while others can. This fact is really easy to overlook, which can lead to all sorts of errors.

2.1 Assignment

To fully understand the range of different types, it's useful to understand variable assignment first.

We've already discussed numbers and functions that apply to numbers. For example, we can use all the usual mathematical operators:

```
>>> 3 * 4
12
>>> 7 - 15
-8
```

We can also store – and later recall – any value. We refer to this as *variable assignment*, or *variable binding*; we take some value and put it into a named location or receptacle. The syntax is simple: the name occurs on the left, the value on the right, and = goes in between. For example:

```
>>> x = 17
>>> bananas = 3.4
```

Note that variable assignment is *not* the same thing as equality. The equals sign does not assert that bananas and 3.4 are the same thing; we're taking the value 3.4 and giving it the name bananas. Remember that the name is always to the left of = and the value is always to the right.[1]

Once a variable name has been bound to a value, the variable can be used anywhere a value of that type can be used. For example:

```
>>> 3 + x
20
>>> bananas * 7
23.8
```

Variable names must begin with a letter or an underline. The remainder of the name can consist of any additional letters, underlines, or numbers. Variable names are case-sensitive and should not contain any of the reserved Python words:

```
False    class     finally  is        return
None     continue  for      lambda    try
True     def       from     nonlocal  while
and      del       global   not       with
as       elif      if       or        yield
assert   else      import   pass
break    except    in       raise
```

These words have specific interpretations in Python.

The right side of an assignment can include functions and variables. For example:

```
>>> oranges = bananas * 4
>>> bananas = bananas - 5
```

Notice that the latter statement confirms that even though assignment uses the symbol =, it is not to be confused with some sort of equality test. For how could bananas be equal to itself minus 5? What this statement does is subtract 5 from whatever value bananas currently has and assign the resulting new value to bananas. If we instead want to test for actual equality, we use ==, as exemplified below:

```
>>> 2 + 2 == 4
True
```

[1] We treat equality operators like += later starting on page 40.

```
>>> 'hats' == 'hat' + 's'
True
>>> bananas == bananas - 5
False
```

2.2 Variable Names

Once we have covered control structures, we'll see that variables are essential in programming. For now, note that the choice of variable name is extremely important. While in principle we can name our variables in many different ways, we should always give our variables a name that reminds us what kind of value we are storing in the variable. For example, if we had a variable that stores somebody's age, a very convenient name would be something like `age`. Far less useful would be a name like `theNumber`. While the latter does imply that the variable contains a number, it does nothing to help us remember *what* number we have stored. For the same reason, variable names like `a` are rather useless.

At the beginning stages of programming, you will be tempted to name your variables cryptically, e.g., `a`, or whimsically, e.g., `ernie` (a recent beloved pet), but make an effort to avoid this and develop good habits early. The best practice is to employ variable names that remind you of the variable's contents.

The best variable names are unambiguous. For example, if we need a variable to store the age of a student, something like `ageOfTheStudent` would be perfect. The problem is that fully explicit variable names are long and perhaps annoying to type out every time the variable is invoked. It's human nature to gravitate to shorter and less descriptive names.

A good rule of thumb is that variable names should be more explicit when they need to be. A variable that lives on through a program that runs many lines should probably have a name that describes the data that variable contains. If at some juncture of a program there are many variables active at the same time, then again, their names should be more descriptive.

2.3 Basic Data Types

We've already discussed basic data types in the previous chapter. Let's look a little more closely at them.

2.3.1 Numbers

There are three basic types of numbers: integer, floating point, and complex. (As linguists, we're only going to be using the first two, so we focus on them.) These are all represented in the usual way (though the suffix for complex numbers is `j`, not `i`). There are functions for converting numbers to and from strings as well.

Type	**Example**	**Convert from string**
integer	`3`	`int('3')`
float	`7.2`	`float('7.2')`
complex	`2+4j`	`complex('2+4j')`

All of these types of numbers can be converted to strings with the `string()` function.

2.3.2 *Booleans*

There is also a basic data type for *booleans*, expressions that are true or false. This class includes only two basic members: `True` and `False`. These can be manipulated with the basic logical operators: `and`, `or`, and `not`. We can combine these operators to construct simple logical expressions:

```
>>> True and False
False
>>> not (False or False)
True
>>> not (not True)
True
```

Here and elsewhere, we use parentheses to clarify the scope of operations. The following are *not* equivalent.

```
>>> not (False or True)
False
>>> (not False) or True
True
```

We can also assign boolean values to variables and construct expressions over them:

```
>>> x = True
>>> y = False
>>> x and y
False
>>> (y or (not y))
True
```

We can also do comparisons on other data types that result in boolean values. For example, we have the following numerical comparisons:

Comparison	**Example**
equality	`3 == 2 + 1`
inequality	`7 != 2`
greater than	`5 > 3`
greater than or equal	`5 >= 5`
less than	`2 < 7`
less than or equal	`5 + 6 <= 10 * 3`

These same operations can be used with strings, although with different interpretations.

Comparison	**Example**
equality	`'hat' == 'hat'`
inequality	`'hat' != 'cat'`
follows alphabetically	`'hat' > 'cat'`
follows alphabetically or equal	`'Hat' >= 'hat'`
precedes alphabetically	`'Hat' < 'hat'`
precedes alphabetically or equal	`'hat' + 's' <= 'chair'`

We can now build up fairly complex expressions. For example:

```
>>> len('hat') > 7 - 2 or 7 - 2 == 3
False
>>> True and (8 = 4 + 4) and (True or False)
True
```

We've already seen that the `type()` function returns the type of an element. For example:

```
>>> type(3)
<class 'int'>
>>> type('3')
<class 'str'>
```

There is a boolean version of this as well, `isinstance()`, that we can use to test if some element or variable is of a particular type. For example:

```
>>> isinstance(3,int)
True
>>> isinstance(3.0,float)
True
>>> isinstance('hat',str)
True
>>> isinstance(3 == 4,bool)
True
```

```
>>> isinstance(3 + 4j,complex)
True
```

2.3.3 *Strings*

For linguists, strings are an extremely important data type. Linguists are concerned with sounds, words, and sentences, and these are typically most naturally represented as strings of characters.

Python offers what looks like two different kinds of strings. First, we have simple strings marked with single or double quotes. They are interchangeable:

```
>>> 'hat' == "hat"
True
```

A second type of string is a triple-quoted string, marked with three single or double quotes. Triple-quoted strings that employ single quotes allow a string to continue over multiple lines.

```
'''This is a string
that continues on
more than one line'''
```

You can use double quotes as well:

```
"""This is a string
that continues on
more than one line"""
```

Triple-quoted strings have slightly different behavior than the other types of strings. If you assign such a string to a variable and then type the variable name, it looks like a single line:

```
>>> x = '''This is
more
than one line'''
>>> x
'this is\nmore\nthan one line'
```

But that's not exactly what it is. It's true that a triple-quoted string is really the same as any other string; it differs only in that it allows the string to be defined over multiple lines. It is not the case, however, that the string above is a single line. The character `\n` represents a line break and will display as such when the variable is explicitly printed:

```
>>> print(x)
This is
more
than one line
```

In fact, you can use single or double quotes to specify a multiline string if you enter `\n` directly:

```
>>> y ='this is more\nthan one\nline too'
>>> print(y)
this is more
than one
line too
```

Thus the way to think about triple-quoted strings is not that they are a different kind of string, but rather a convenient way to enter multiline strings.

A similar issue arises with other special characters, e.g., tab. One cannot enter the tab character directly in either of the string types we've discussed in the interactive environment. Instead, a tab can be typed into either type of string as `\t`. Notice how the tab is not displayed properly when the name of the variable is typed, but only when the variable is given as an argument to `print()`.

```
>>> x = 'a\tfew\ttabs\there'
>>> x
'a\tfew\ttabs\there'
>>> print(x)
a	few	tabs	here
```

We've now got two cases where backslashes are required to type a special character, tab or newline, into strings: `\t` and `\n`. But what if you actually want to type a backslash followed by a `t` or an `n`? With a single-, double-, or triple-quoted string, you have to enter a double backslash `\\`. If you assign the string to a variable name and then type that name, you see the double slash; but if you invoke the variable with `print()`, then only a single slash is displayed:

```
>>> x = 'xyz\\nabc'
>>> x
'xyz\\nabc'
>>> print(x)
xyz\nabc
```

If you're entering strings with lots of special characters that require backslashes, the *raw* string notation becomes useful. If you prefix a single- or double-quoted string with `r`, then a backslash is interpreted as an actual backslash. Thus:

```
>>> x = r'abc\nxyz'
>>> y = r"xyz\nabc"
>>> x
'abc\\nxyz'
>>> print(x)
abc\nxyz
>>> y
'xyz\\nabc'
>>> print(y)
xyz\nabc
```

Keep in mind that the *raw* string notation does not create a different kind of string; rather it is a way of entering the string differently.

We have already seen functions and operators that apply to strings:

```
>>> x = '77'
>>> len(x)
2
>>> int(x)
77
```

We've also seen that there are *methods* that apply specifically to strings. Recall that there is a different syntax for methods. For example:

```
>>> x = 'a Hat'
>>> x
'a Hat'
>>> x.count('a')
2
>>> x.upper()
'A HAT'
>>> x.lower()
'a hat'
```

An extremely useful method for strings is `format()`. The basic idea is that you define a string with empty slots and then fill those slots later with the `format()` method. Slots are indicated with `{}`. So we can define a suitable string with one slot like this:

```
>>> x = '{} Mike'
```

We can then fill that slot any number of ways:

```
>>> y = x.format('Hello')
>>> y
'Hello Mike'
>>> z = x.format('Goodbye')
>>> z
'Goodbye Mike'
```

A string can have multiple slots. To fill them, the `format()` method can take multiple arguments.

```
>>> x = '{} Mike. {}?'
>>> x.format('Hello','How are you')
'Hello Mike. How are you?'
>>> x.format('Hwyl','Sut wyt ti')
'Hwyl Mike. Sut wyt ti?'
```

When a string does have multiple slots, they can be numbered (from 0). They are then filled in that order by the arguments to `format()`.

```
>>> x = 'one = {1}; two = {0}'
>>> x.format('dos','uno')
'one = uno; two = dos'
```

In fact, the slots can be named and specified in the call to `format()`.

```
>>> x = 'one = {uno}; two = {dos}'
>>> x.format(dos='dau',uno='un')
'one = un; two = dau'
```

Notice how the arguments to `format()` can come in either order since the naming disambiguates. The following code snippet has the same effect as the preceding.

```
>>> x.format(uno='un',dos='dau')
'one = un; two = dau'
```

One can also do simple spacing and alignment with the string method `format()`. You indicate the number of spaces to reserve for the argument with a preceding colon. If the argument supplied by `format()` is less than the space reserved, you indicate alignment to the left, right, or center with <, >, and ^ respectively.

```
>>> x = '-{:<10}-'
>>> x.format('hat')
'-hat       -'
>>> x = '-{:>10}-'
>>> x.format('hat')
'-       hat-'
>>> x = '-{:^10}-'
>>> x.format('hat')
'-   hat    -'
```

Notice that when there is an odd number of remaining spaces, the extra space goes to the right. There is a lot more that can be done with the `format()` method, but we set it aside for now.

Finally, there is a syntax for extracting part of a string. The syntax is to put one or more integers in square brackets after the string. Using a single integer in the square brackets extracts a single character. The characters in a string are counted from the left, starting at 0.

h	a	p	p	i	n	e	s	s
0	1	2	3	4	5	6	7	8

For example:

```
>>> x = 'happiness'
>>> x[1]
'a'
>>> x[5]
'n'
```

You can also refer to a sequence of characters by using two integers separated by a colon. The first index is the starting point of the sequence; the second index is just *after* the end of the sequence. For example, `[2:4]` starts at the item with index `2` and extends to just before the item with index `4`, thus the item with index `3`. Since indexing starts at `0`, this span goes from the third character in the string to the fourth character.

```
>>> x = 'abcde'
>>> x[2:4]
'cd'
```

Thus `x[n:n+1]` is the same as `x[n]`. If we have `x[n:m]` where `m < n+1`, we end up with an empty string. For example:

```
>>> x = 'abcde'
>>> x[2:2]
''
>>> x[2:1]
''
```

Interestingly, we can leave out either integer when the colon is present. Leaving out the second index gives the entire string starting from the first integer present. Leaving out the first index returns the entire string up to just before the integer that's present. For example:

```
>>> x = 'abcde'
>>> x[2:]
'cde'
>>> x[:2]
'ab'
```

There are many more things we can do with strings, but these are the most critical.

2.3.4 *Lists*

Lists are an extremely important data type: a single structure that holds a sequence of elements of whatever type you want. You can create a list by simply listing elements in square brackets. For example:

```
>>> x = [1,6,4,9]
>>> y = ['stops','fricatives','glides']
>>> z = [7,'hats',56,'chairs',6.802]
```

The `len()` command applies to lists:

```
>>> len(x)
4
>>> len(y)
3
>>> len(z)
5
```

You can also index into a list, as with a string:

```
>>> x[1]
6
>>> y[0]
'stops'
>>> z[3:]
['chairs',6.802]
```

There are lots of methods for dealing with lists. For example, the `append()` method adds an element at the end of the list:

```
>>> x = ['rocks','paper']
>>> x
['rocks','paper']
>>> x.append('scissors')
>>> x
['rocks', 'paper', 'scissors']
```

The `pop()` method removes an element at a specified index position in a list. The method then returns that element, altering the list at the same time.

```
>>> x = ['stops','fricatives','glides']
>>> x.pop(1)
'fricatives'
>>> x
['stops','glides']
```

The mirror-image method is `insert()`, which takes two arguments, the index and the element to insert. The element is inserted just *before* the index you specify.

```
>>> x = ['stops','fricatives','glides']
>>> x.insert(1,'hello!')
>>> x
['stops','hello!','fricatives','glides']
```

A function that will turn out to be quite useful later on is `range()`. This takes a single integer argument and returns a `range` object that represents the sequence from `0` up to that argument. This can be directly converted to a list by using the `list()` function. For example:

```
>>> x = list(range(4))
>>> x
[0,1,2,3]
```

We also have the `sort()` and `reverse()` methods. The first sorts a list, and the second reverses it. Note that `sort()` works only for a list of uniform objects that are, in fact, sortable.

```
>>> x = [5,2,8,3]
>>> x.sort()
>>> x
[2,3,5,8]
>>> x = [5,2,8,3]
>>> x.reverse()
>>> x
[3,8,2,5]
```

Finally, we have the `in` operator, which we can use to test for membership in a list:

```
>>> x = [5,2,8,3]
>>> 8 in x
True
>>> 7 in x
False
```

2.3.5 *Tuples*

Another data type that you will see quite often is a *tuple*, a fixed sequence of elements that is similar to a list in many ways. The key difference is that tuples are fixed in length and, once created, cannot be changed. (We will discuss this notion more in Section 2.4 below.)

You create a tuple with parentheses. An empty tuple is just parentheses: `()`. Larger tuples separate the members with commas, e.g., `(7,'hat',8.2)`. Interestingly, a tuple with one element must have a comma, unlike a list with a single element: `(3,)` vs. `[3]`.

```
>>> x = ()
>>> y = (7,'hat',8.2)
>>> z = (3,)
```

The `len()` function applies to tuples, and you can index tuples just like lists.

```
>>> y = (7,'hat',8.2)
>>> len(y)
3
>>> y[2]
8.2
```

The `in` operator applies to tuples, just as it does to lists:

```
>>> x = (5,2,8,3)
>>> 8 in x
True
>>> 7 in x
False
```

Finally, we can convert a list to a tuple with `tuple()`, or a tuple to a list with `list()`:

```
>>> x = [1,2,3]
>>> type(x)
<class 'list'>
>>> y = tuple(x)
>>> type(y)
<class 'tuple'>

>>> a = (1,2,3)
>>> type(a)
<class 'tuple'>
>>> b = list(a)
>>> type(b)
<class 'list'>
```

2.3.6 *Dictionaries*

One of the most useful data types for linguists is *dictionaries*. Dictionaries are effectively sets of pairs, where the first element in each pair can be used to "look up" the second element. The first element of each pair must thus be distinct from the first elements of all the other pairs. A dictionary is marked with curly brackets within which each pair of elements is separated with a colon. For example:

```
>>> d = {'cat':7,'chair':'hat','table':7}
```

Notice how `'cat'`, `'chair'`, and `'table'` are distinct, but `7`, `'hat'`, and `7` are not, and do not need to be. We refer to the first member of each pair as a *key* and the second as its *value*. We look up values by putting the key in square brackets after the name of the dictionary:

```
>>> d = {'cat':7,'chair':'hat','table':7}
>>> d['cat']
7
>>> d['chair']
'hat'
```

The `len()` function applies to a dictionary and returns the number of pairs in the dictionary:

```
>>> d = {'cat':7,'chair':'hat','table':7}
>>> len(d)
3
```

We can add new pairs to a dictionary simply by assigning to a new key:

```
>>> d = {'cat':7,'chair':'hat','table':7}
>>> d['onion'] = 3.7
>>> len(d)
4
```

We can test for whether any specific item is a key in a dictionary with `in`. For example:

```
>>> d = {'cat':7,'chair':'hat','table':7}
>>> 'chair' in d
True
>>> 'hat' in d
False
```

Notice how `in` tests for membership in the dictionary keys, not the dictionary values.

Dictionary items can be altered or deleted.

```
>>> d = {'cat':7,'chair':'hat','table':7}
>>> d['cat'] = d['cat'] + 2
>>> d['cat']
9
>>> del(d['cat'])
```

```
>>> d
{'chair':'hat','table':7}
```

We can extract the keys, values, or key–value pairs from a dictionary. These are returned in a specific format we've not discussed yet, but all can be converted to lists by using `list()`. For example:

```
>>> d = {'cat':7,'chair':'hat','table':7}
>>> list(d.keys())
['chair', 'cat', 'table']
>>> list(d.values())
['hat', 7, 7]
>>> list(d.items())
[('chair', 'hat'), ('cat', 7), ('table', 7)]
```

Finally, dictionaries can be used directly with the string method `format()`. There's a special operator for this `**`. (Remember, however, that this means exponentiation in a numerical context.) The basic idea is that the slots in the string are named. The values that go in those slots are associated with the relevant keys of the dictionary. For example:

```
>>> d = {'uno':'eins','dos':'zwei',
   'tres':'drei'}
>>> s = 'one = {uno} and three = {tres}'
>>> s.format(**d)
'one = eins and three = drei'
```

Notice that keys of the dictionary that are not named in the string are simply not used.

2.4 Mutability

An extremely important and, unfortunately, confusing concept is *mutability*. Some objects in Python can be changed and others cannot. To understand this concept fully though, we must understand a couple other notions: *garbage collection* and *naming*.

Because this subject is potentially confusing and challenging, work through this section carefully. You may need to return here later as you learn more and can contextualize this information better.

An element created in Python occupies space in your computer's memory. As you create more elements, that memory fills. As you interact with Python, or as your Python programs run, you run the risk of filling up your computer's memory. Python manages this memory for you automatically. As you create and manipulate elements, Python holds them in memory. When you cease to use an element, Python detects this and removes the element, freeing up memory. This latter process is called *garbage collection.*

One way garbage collection applies is to named elements that are no longer used; for example, if you define `i` as `35` at some point in a program, use it, and then stop using it. The garbage collector will detect that and free up that bit of memory.

Another instance where garbage collection applies is in a situation like the following:

```
>>> i = 7
...
>>> i = 'hat'
```

Here we have created an integer and named it `i`. Later, we create a string `'hat'` and name it `i`. The integer is now, in principle, floating around unnamed. The garbage collector detects this and frees up that memory accordingly.

In this case, it's important to be clear that `i` is not some data element that has changed. Rather, `i` is the name of an integer, and that name is reassigned to a string. The integer element becomes available for garbage collection. Memory is allocated for the new string, and it takes on the name `i`.

In this light, let's now consider *mutability*. Lists and dictionaries are mutable elements that can be directly changed. Everything else we've discussed, integers, floating point numbers, complex numbers, booleans, strings, and tuples, are immutable, elements that cannot be changed.

Given our discussion of memory allocation with respect to the example given above, mutability can be detected only when an element has multiple parts. The idea would be that you could change some part of an element and leave the rest intact. This is possible for lists and dictionaries; they are mutable. Thus, we can define a list and then add or delete elements in that list, or change elements in the list. For example:

```
>>> x = ['Tom','Dick','Harry']
>>> x[1] = 'Mary'
>>> x
['Tom','Mary','Harry']
>>> x.append('Edna')
>>> x
['Tom','Mary','Harry','Edna']
```

We've already seen the list methods `insert()` and `pop()` in Section 2.3.4 above, which both allow us to change a list.

Dictionaries, as mentioned, also are mutable. We can add, delete, or change items. For example:

```
>>> d = {'un':'un','deux':'?','trois':'tri'}
>>> d['quatre'] = 'pedwar'
>>> d['deux'] = 'dau'
>>> d
{'un':'un','deux':'dau','trois':'tri',
  'quatre':'pedwar'}
```

As noted above, for simple data types like numbers and booleans, we cannot see the effects of their immutability. Superficially, it seems like we can change them:

```
>>> x = 3
>>> x = 7
>>> x
7
>>> y = True
>>> y = False
>>> y
False
```

This is only apparent, however. In both cases above, we are simply reassigning the names `x` and `y` to new elements. The old elements are unchanged, though available for garbage collection.

Strings and tuples have multiple parts, but they are *not* mutable. For example, while we can refer to pieces of each with indexes, we cannot assign new values to just those indexed segments. For strings:

```
>>> x = 'abc'
>>> x[1] = 'd'
  Traceback (most recent call last):
  File "<stdin>", line 1, in <module>
  TypeError: 'str' object does not
  support item assignment
```

The same applies to tuples:

```
>>> x = ('a','b','c')
>>> x[1] = 'd'
  Traceback (most recent call last):
  File "<stdin>", line 1, in <module>
```

```
TypeError: 'tuple' object does not
support item assignment
```

Notice that we can reuse a name that initially was assigned to a string or tuple, but this is reusing the name, not changing the initial data structure.

```
>>> x = 'a first string'
>>> x = 'another string'
```

Notice that methods for mutable and immutable elements differ in whether they change the objects of which they are methods or return a new element. For example, the `reverse()` method for lists alters the list in place, reversing its elements. On the other hand, the `upper()` method for strings returns a new string with all uppercase letters *and leaves the initial string unaffected.*

```
>>> x = [1,2,3]
>>> x.reverse()
>>> x
[3,2,1]
>>> y = 'letters'
>>> y.upper()
'LETTERS'
>>> y
'letters'
```

Mutability may seem like a fine point, but try to keep it in mind as you proceed. It's extremely easy to forget, and neglecting it can lead to all sorts of hidden bugs in your programs.

2.5 Exercises

2.1 What does this code do: `isinstance(bool == bool,bool)`? What is the output and why?

2.2 What does this code do: `a = 3 == 3`? What is the result and why? What's the difference between that code snippet and this one: `a == 3 = 3`? (Note that the latter produces an error.)

2.3 Write three lines of code that: (i) create a variable with the value 37; (ii) create another variable with the value 4; (iii) print out the result of multiplying those together.

2.4 Write three lines of code that: (i) create a string *transformational*; (ii) create a string *grammar*; (iii) concatenate those together and print the result. (Make sure there is a space between the two words!)

2.5 Explain what the following bit of code does and why it's a bad idea:

```
yes = 'no'
no = 'yes'
```

2.6 Write a single line of code that tests whether 7^3 is greater than 15×16.

2.7 Explain what's going on below; is x mutable?

```
x = 'abc'
x = x.upper()
```

2.8 Semitic morphology involves intercalating vowels and consonants to express morphological categories. For example, the Arabic root k,t,b occurs in the following forms:

katab-a	'he wrote'
kaatab-a	'he corresponded'
kutib-a	'it was written'
kitab	'book'
kuttaab	'writers'
uktub	'write!'
etc.	

How might you use `format()` to describe this system? Give a sample representation for the root k,t,b and show how `format()` could be used to express different categories.

2.9 We can also use `format()` to treat vowel harmony. Choose a simple vowel harmony system and show how this might work.

2.10 You can use the dictionary data type in conjunction with the `format()` command to do translations between two languages. Choose two languages, construct a small dictionary for them, then create some strings with which you can use your dictionary to do translations.

2.11 Lists and strings are similar in many ways. Describe and show with code three ways in which they are *not* alike.

2.12 **Web**: Snoop around on the web and figure out how the `set` data type works. Explain and exemplify.

3 Control Structures

We've learned a fair amount of Python, but our code so far is unsatisfying. Basically, our code at this point can't really do any more than what we can type in. There is no *savings*, no extra bang for our buck. This changes in the current and following chapters, where we discuss control structures and then input–output.

In previous chapters a program is just a sequence of statements, executed one after the other. In this chapter, we show how statements can be executed conditionally, how some statements can escape being executed, and how others can be executed any number of times. In effect, this allows for much more effect than "what you type in." You'll see that we can do an infinite amount of stuff with a finite amount of code. In principle, the control structures we will learn are simple, but – along with variable assignment (Section 2.1) – they are the heart and soul of a programming system, almost everything you need to know to write meaningful and useful programs.

From here on, we will assume that all code is written as a program; that is, the code is typed into a file, and then that file is run as a program. All programs in this book are available on the book website, and the relevant filename is given below each listing.

3.1 Grouping and Indentation

To make full use of control structures, we need to understand how Python groups statements together. To understand this, let's look at the `if` control structure. This structure, in its simplest form, allows for conditional application. There is a test clause and then a block of one or more statements. If the test clause is true, the statements are executed. If it is not true, the statements are not executed. Here's an example:

```
#single statement on the same line
if 2+2 == 4: print('that was true')
```

control1.py

or:

```
#single statement indented on next line
if 2+2 == 4:
   print('that was true')
```

control2.py

The test is introduced with the keyword `if` and terminates with a colon. If the statement block is a single statement, it can occur on the same line or the next. If it occurs on the next line, it must be spaced or tabbed in at least one space or tab. If there is no indentation, an error is produced. The example below demonstrates. Note that we have added a comment to the code to remind us that this file does not work. To remind you, comments are indicated with a # on the left side of the line and have no effect on the code. If you run this or any other program, comments have no effect.

```
if 2+2 == 4:
print('that was true')  #produces an error
```

control3.py

If there are multiple statements contingent on the test, then they must appear on separate lines *and* they must be tabbed or spaced with *the same number of spaces or tabs*. Here's a well-formed example:

```
#multiple statements correctly indented
if 2+2 == 4:
   print('that was true')
   print('...really true')
```

control4.py

Here's an ill-formed example; the second statement has an extra space.

```
if 2+2 == 4:
   print('that was true')
    print('...really true')   #produces error
```

control5.py

A similar error occurs if the first of the two statements has extra indentation:

```
if 2+2 == 4:
    print('that was true')   #produces error
   print('...really true')
```

control6.py

In principle, you can indent with either spaces or tabs, but *be careful and do not mix them!* For example, if you indent one line with a tab and then indent the next line with the right number of spaces so that they *look* aligned, this will typically generate an error. The problem is that Python has no way of knowing how wide you've set your tabs to display in your word processor.

Note that this indentation has semantic consequences. Compare the following two programs:

```
if 2 + 2 == 5:   #2-statement block
   print("that shouldn't happen")
   print('or this....')
```

control7.py

```
if 2 + 2 == 5:   #1-statement block
   print("that shouldn't happen")
print('or this....')
```

control8.py

In both cases, we test if `2 + 2 == 5`; that evaluates to false. In the first example, the two statements are not executed since they are in the block of contingent statements. In the second case, the second statement *is* executed. Since it is not indented, it is evaluated as being outside the `if` structure and thus not contingent on the `if` test.

This semantic effect also shows itself with nested `if` structures. In the following example, we first test if `2+2==4`. If that is true, which of course it is, then we execute the following three statements. We first print `'wow'`. Second, we execute another `if` structure. This second if structure tests if `7*7==48`, which is false. If it were true, we would print `'wow again'`. Finally, the third statement in the initial `if` structure's block prints `'wow a third time'`.

```
#nested if with separate statement
if 2+2==4:
   print('wow')
   if 7*7==48:
```

```
        print('wow again')
    print('wow a third time')
```

control9.py

In the next example, we've spaced/tabbed in the final statement so that it is part of the nested `if` structure. It will execute only if *both* `if` tests evaluate to true.

```
#nested if with two-statement block
if 2+2==4:
    print('wow')
    if 7*7==48:
        print('wow again')
        print('wow a third time')
```

control10.py

Finally, in this third case, the final statement has no indentation. It therefore is executed regardless of the two tests.

```
#nested if, with independent statement
if 2+2==4:
    print('wow')
    if 7*7==48:
        print('wow again')
print('wow a third time')
```

control11.py

3.2 `if`

Let's now look a little more closely at the options for `if`. We've already seen that one or more statements can be contingent on some logical test. If the test evaluates to true, the block of statements then executes in order. If it is false, none of them is executed.

A simple augmentation of the `if` structure is the `else` clause. This is a block of code that executes if the `if` test evaluates to false. The `else` statement follows the first block of code and is at the same level of indentation as the initial `if` clause. For example:

```
if 2+2==5:    #only else block executes
    print("this won't print")
```

```
else:
    print('but this will')
    print('...and so will this')
```

control12.py

If the `if` clause evaluates to true, the `else` block will not execute:

```
if 2+2==4:    #else-block does not execute
    print('this will print')
else:
    print("but this won't")
    print('...and neither will this')
```

control13.py

Finally, one can add any number of `elif` clauses to an `if` structure. These add additional *contingent* tests to the structure. The syntax is like this:

```
if test1
    block1
elif test2
    block2
elif test3
    block3
...
else
    blockn
```

We can represent graphically which blocks in the example just above are executed depending on which tests evaluate to true:

test1	test2	test3	what applies?
True	True	True	block1
True	True	False	block1
True	False	True	block1
True	False	False	block1
False	True	True	block2
False	True	False	block2
False	False	True	block3
False	False	False	blockn

Notice how if test1 is true, block1 applies regardless of the truth or falsity of any other test. If test1 is false and test2 is true, block2 applies regardless of whether test3 is true. Finally, the else block applies only if all previous tests are false.

A final complication is that it may sometimes be convenient to have an empty block. Imagine you want to test a string for whether the first letter is 'a', but then do something in all other cases. One way to do this is to test for that, but then put your code in the else block. The problem with this is that you would then have an empty block following the if test, which is not allowed in Python. To deal with this possibility, Python has the pass statement, which you can use to fill an empty block. For example:

```
x = 'hat'     #set a variable
if x[0] == 'a':
   pass       #if clause with empty block
else:
   print('doing something here....')
```

control14.py

Here the print statement is executed only when the first letter of the string is *not* 'a'.

3.3 Digression on Printing

We will spend more time on this topic in Chapter 4, but let's add a little detail to how we can explicitly print something in Python using the print() command.

We've already seen that print() can be given a string as an argument and displays that string on the screen:

```
>>> print('here is a string')
here is a string
```

It can also be given any number of strings; they are all printed to the screen:

```
>>> print('one','two','three')
one two three
```

Notice that the strings are separated by spaces. We can specify another separator or, indeed, no separator with the optional sep argument. If we include this argument, we specify its value with =. For example:

```
>>> print('one','two','three',sep='-')
one-two-three
>>> print('one','two','three',sep='')
onetwothree
```

Finally, notice that the default behavior for `print()` is to print its argument(s) and begin a new line. We can specify different behavior by giving a value to the optional `end` argument.

```
>>> print('one','two',sep='-',end='!')
one-two!>>>
```

We will exploit these options in the exemplification of the control structures to come.

3.4 for

The `for` control structure allows for multiple application of some fixed block of code. You specify a list or sequence of items and then iterate over the list, applying the block once for each item in the list or sequence. The syntax is a `for` clause followed by an indented block of one or more statements. The `for` clause begins with the keyword `for`, then a variable name, then the `in` operator, then a list or sequence of items that can be iterated over, and finally a colon.

Here's a simple example:

```
for i in [1,2,3]:
   print(i)
```

control15.py

Here we assign the variable `i` the values from the list `[1,2,3]` one by one. For each assignment, we apply the block. In this case, the block simply prints the value of `i`:

```
1
2
3
```

Note that nothing requires that we actually use the value `i` in the block. Thus the following works as well:

```
for i in [1,2,3]:
   print('wow')     #value for i not used
```

control16.py

The variable `i` again assumes each value in the list, but does nothing with them. This produces:

```
wow
wow
wow
```

Similarly, nothing prevents us from using the variable more than once per each iteration. For example:

```
for i in [1,2,3]:     #using variable twice
   print('{} + 2 = {}'.format(i,i+2))
```

control17.py

We can iterate over strings as well. For example:

```
for i in 'tone':
   print(i,end=' ')     #letters with spaces
print()                 #add return at end
```

control18.py

Here we print each letter separately. We end each print operation with a space rather than a return. Then, after all that printing, we get back to the beginning of the line by printing nothing. Since the default is to end each printed item with a return, printing nothing has the effect of starting a new line.

```
t o n e
```

Recall the `range()` function from Section 2.3.4 above. We can iterate directly over the sequence of numbers it provides. For example, if we want to add up all the numbers from `0` to `4`, we can do it like this:

```
total = 0               #set variable
for i in range(5):      #iterate...
```

```
    total = total + i #...add to total
print(total)          #print total
```

control19.py

Here we first define a variable `total` and set its value to `0`. We then create a sequence from `0` to `4` (the integer value just before the value specified). Then on each iteration we reset the value of `total` to be equal to its current value plus the current value of `i`.

Let's chart this out to see how it works. First, we number the lines above so we can refer to where we are:

```
1 total = 0
2 for i in range(5):
3     total = total + i
4 print(total)
```

Now we step through the code line by line, showing the values of `total` and `i` at each point.

Line	`total`	`i`
1	0	NA
2	0	0
3	0	0
2	0	1
3	1	1
2	1	2
3	3	2
2	3	3
3	6	3
2	6	4
3	10	4
4	10	4

Notice how we go back and forth between lines 2 and 3 alternately resetting the values for `i` and `total`.

A fairly common programming technique is to create a variable that we incrementally change in some sort of loop. We can do some fairly interesting stuff with this. For example, we might do morphological recursion as follows:

```
prefix = 'anti'                #initial prefix
word = 'missile'               #initial word
print(word)                    #print word
```

```
for i in range(3):                 #iterate 3 times
   word = prefix + '-' + word #add prefix...
   print(word)                     #print new word
```

control20.py

This produces:

```
missile
anti-missile
anti-anti-missile
anti-anti-anti-missile
```

There are a couple of things that distinguish this code from the previous example. First, notice that we are doing string concatenation rather than addition. Second, notice that we are printing the `word` variable at each iteration rather than just at the end. This, then, is an instance of the semantic effects of indentation in a real example. If we had not indented the final print statement, this code would have simply printed the last line above.

As with `if`, we can nest `for` structures. Thus, we can augment our prefixation example above so that it prefixes multiple words:

```
#define prefix and 3 words
prefix = 'anti'
words = ['missile','racism','music']
for word in words:    #iterate over words
   print(word)          #print word
   #for each word, iterate 3 times
   for i in range(3):
      #add prefix to current word
      word = prefix + '-' + word
      print(word)     #print new word
```

control21.py

Here we go through the words one by one. We prefix each word three times. Here's the output:

```
missile
anti-missile
anti-anti-missile
anti-anti-anti-missile
```

```
racism
anti-racism
anti-anti-racism
anti-anti-anti-racism
music
anti-music
anti-anti-music
anti-anti-anti-music
```

The other thing to notice here is that `for` allows us to do potentially massive calculations with finite means. For example, if we wanted to sum numbers up to 10,000, this would involve a trivial change to the code above.

```
#variable to accumulate additions
total = 0
for i in range(10000):   #iterate a lot
   total = total + i     #add i to total
print(total)             #print total
```

control22.py

A word to the wise at this point. Imagine you do want to do some massive looping operation, maybe even with lots of nested loops. A really good rule of thumb here is to try out such operations *first* with a much smaller number of loops, make sure everything behaves as expected (the math works out, the right things are printed, etc.), *then* rewrite the code with the larger iterations.

Now that we have two control structures, `if` and `for`, we can combine them. Let's write some code to count the vowels in a string.

```
#define vowels
vowels = 'aeiou'
#variable to accumulate count
vowelCount = 0
word = 'Appalachicola'                #define word
#go letter by letter
for letter in word:
   if letter in vowels:               #vowel?
      vowelCount = vowelCount + 1 #add 1
print(vowelCount)                     #print total
```

control23.py

Here we first define vowels. We then set our count of vowels to 0. We define the word we will count vowels in. Next we iterate through each letter in the word assigning the current letter to the variable `letter`. We then test if that letter is a vowel with the `in` operator. If it is, we augment the value of `vowelCount` by one. If it's not a vowel, that test fails, nothing happens, and we go on to the next iteration. Finally, we print the value of `vowelCount`.

The operation where we augment a variable by some specific amount is frequent enough that there is a special operator for it: +=. The following code does exactly the same thing as the preceding.

```
vowels = 'aeiou'             #define vowels
#variable to accumulate count
vowelCount = 0
word = 'Appalachicola'       #define the word
for letter in word:          #go letter by letter
    if letter in vowels:     #check if vowel
        vowelCount += 1      #if so, add 1
print(vowelCount)            #print the total
```

control24.py

There are, in fact, similar equality operators for other simple operators, e.g., -=, *=, /=.

We can also nest a `for` loop inside an `if` structure. Here we set up virtually the same initial variables. We then test if the first letter of the word is a vowel. Note that since we've defined vowels as lowercase letters, we must first convert the first letter to lowercase. If the word does begin with a vowel, we count all the letters. (We could have simply used the `len()` function instead here!)

```
vowels = 'aeiou'             #define vowels
letterCount = 0              #set counter to zero
word = 'Appalachicola'       #define word
#is 1st letter a lowercase vowel?
if word[0].lower() in vowels:
    for letter in word:      #go letter by letter
        letterCount += 1     #for each letter, add 1
#do this if word is not V-initial
else:
    print('Not vowel-initial')
print(letterCount)           #print letter total
```

control25.py

Make sure you understand the semantics of control structures like `if` and `for` and what it means for them to be nested in different ways. *Writing programs is mostly about figuring out the logic of the problem you want to solve and then recasting it in terms of nested control structures.*

3.5 while

Technically, the `for` control structure isn't necessary. Anything you can do with `for`, you can do with `while`, perhaps a bit awkwardly. The logic of `while` is that a statement or block of statements is repeated as long as some test holds true. The syntax is parallel to what we have seen already. There is the keyword `while`, followed by the test (an expression that returns a boolean value), followed by a colon. There is then an indented block of one or more statements. If there is only one statement, it can appear on the same line as the `while` test. Here it is schematically:

```
while x == y:
    dosomething(z)
    dosomethingelse(w)
```

Here we have a test for equality between some variables `x` and `y`. We then have a block of two statements. We first check the test. If it is true, we apply the statements in the block. We then go back and check the test again. If it is still true, we apply the statements again. We go through this loop until the test returns `False`.

This should seem a bit silly. If the test returns `True`, then it should *always* return `True` and we should loop forever. That's not typically what we want! The way this structure is typically used is that the truth value of the test is changed during the block of statements so that it eventually becomes false. Here's a simple example:

```
count = 0              #set counter
while count < 3:       #check counter value
    count += 1         #increment, escape clause!
    print(count)
```

control26.py

First we define a variable `count` and store the number `0` in it. We then begin our `while` structure by testing whether `count` is less than `3`, which of course it is. If we did not alter the value of `count` in the block of statements, this would iterate forever. What we do, though, is increment the value of `count` at each iteration. It will eventually reach 3, the test will then fail, and the iteration will cease.

Again, we can diagram the program flow to see how this goes. First, we number lines:

```
1 count = 0
2 while count < 5:
3     count += 1
4     print(count)
```

Now we can examine the value of `count` at each step:

Line	`count`	**Test**	**Printing**
1	0	NA	
2	0	`True`	
3	1	NA	
4	1	NA	1
2	1	`True`	
3	2	NA	
4	2	NA	2
2	2	`True`	
3	3	NA	
4	3	NA	3
2	3	`False`	

It's important to keep track of what happens when. Small changes can change the outcome. For example, imagine we reverse lines 3 and 4 like this:

```
count = 0               #declare counter
while count < 3:        #check value of counter
    print(count)        #print the value
    count += 1          #NOW increment counter
```

control27.py

With this order, we print the numbers 0, 1, 2 instead.

It's also easy to get this wrong and inadvertently get an infinite loop. For example:

```
count = 0
while count < 3: #error: infinite loop!
```

```
    print(count)
    count = 1
```

control28.py

Here we've replaced `count += 1` with `count = 1`. This means that in every loop, we reset the variable to 1; it will thus *always* be less than 3.

As with the other control structures we have seen, `while` can combine with itself and with other structures. Here is an example of `while` nested within `while`:

```
word = 'alphabet'          #define word
count = 0                  #define counter
#iterate while count is less than word length
while count < len(word):
    print(word[count]) #print current letter
    count += 1             #increment counter
    othercount = 1         #start new counter
    #check if 2nd counter is less than 1st
    while othercount < count+1:
        #print ever larger prefixes
        print('\t',word[0:othercount])
        othercount += 1 #increment other counter
```

control29.py

First we assign the string `'alphabet'` to the variable `word` and the integer `0` to the variable `count`. The outer `while` loop goes through the string letter by letter and prints each one out. It increments the `count` variable on each loop to keep track of the iteration *and* to determine which letter to print. The inner `while` loop is a little more complex. It prints out prefixes (initial substrings) of the word based on the current letter determined by the outer `while` loop. Thus, for example, when the current letter is `'h'`, the inner `while` loops prints out `a`, `al`, `alp`, and `alph`. Here's what the (beginning of the) output looks like:

```
a
        a
l
        a
        al
```

```
p
    a
    al
    alp
h
    a
    al
    alp
    alph
...
```

Notice how the test for the inner `while` loop refers to the variable `count` as well as `othercount`, which change on the outer and inner loops respectively.

The `while` structure can combine with other control structures. For example, here is a case of `if` inside `while`:

```
word = 'alphabet'              #set word
vowels = 'aeiou'               #define vowels
count = 0                      #set counter
while count < len(word):       #iterate
   letter = word[count]        #get current letter
   #is it a non-vowel?
   if letter not in vowels:
      print(letter)            #if so, print it
   count += 1                  #increment counter
```

control30.py

This program collects each letter of the word and checks if it is a vowel. If so, it is printed.

```
l
p
h
b
t
```

The `while` structure is similar to the `for` structure, and it's always possible to translate back and forth. For example, recall the first example of the `for` structure (page 35), repeated below:

```
for i in [1,2,3]:
   print(i)
```

We can translate this into a `while` structure as follows:

```
i = 1
while i < 4:
   print(i)
   i += 1
```

control31.py

The upshot is that the choice between these alternatives typically depends on which makes the code more intelligible to *you* as the programmer.

Finally, we note that the `while` structure also allows for an `else` statement block. This is executed when the `while` test is or becomes false. The general syntax is as follows:

```
while ...:
   statement(s)
else:
   statement(s)
```

Here's a simple example:

```
vowels = 'aeiou'          #define vowels
word = 'Winnepesaukee' #set word
#create two counters
counter = 0
vowelcount = 0
#go letter by letter
while counter < len(word):
   #is current letter a vowel?
   if word[counter] in vowels:
      vowelcount += 1
   #keep track of total number of letters
   counter += 1
#when counter is too big, do this:
else:
   print('There are',vowelcount,
      'vowels in this word')
```

control32.py

Here we go through the word letter by letter, asking whether the current letter is a vowel. If it is, we increment `vowelcount`. When we have reached the end of the word, we execute the `else` clause and print out the number of vowels. Logically, this kind of `else` clause is not necessary. It exists only because some programmers find that it clarifies their logic to use it.

3.6 break and continue

We can impose finer control on `for` and `while` using `break` and `continue` statements. A `break` statement exits from the *smallest* enclosing `for` or `while` loop. The `continue` statement exits from the current iteration of the *smallest* enclosing `for` or `while` loop and moves to the next iteration. Here's an example of `break`:

```
vowels = 'aeiou' #define vowels
word = 'sthenic' #define word
counter = 0         #set up a counter
#iterate through word letter by letter
while counter < len(word):
    #if current letter is vowel, exit loop
    if word[counter] in vowels:
        break
    #don't forget to update the counter!
    counter += 1
#print value of counter when break occurred
print('The word begins with',
    counter,'consonant letters')
```

control33.py

Here we define a specific word and vowel letters. We then check whether each letter is a vowel. If it is, we break/exit from the `while` loop. We then print the value of `counter`, which is the number of letters iterated through to get to the `break` statement.

Note that it's fairly easy to make errors in getting the right count here. We start with `counter` set to 0, which allows us to use it to access each letter of the word, remembering that the indices of the word begin with 0.

We increment `counter` *after* the `if` structure. When the `if` structure test is true, the value of `counter` is the index of the first vowel letter, the index just past the last consonant letter. Since indices start at 0, this means that the value of `counter` when we exit is *also* the total number of consonants in the initial span.

Let's go through this line by line to see. First, we number each line:

```
1 vowels = 'aeiou'
2 word = 'sthenic'
3 counter = 0
4 while counter < len(word):
5    if word[counter] in vowels:
6       break
7    counter += 1
8 print('The word begins with',
9    counter,'consonant letters')
```

Now we step through the code, keeping track of the values of `counter` and `word[counter]`. We collapse sequences of steps together if the relevant values don't change.

Line	`counter`	`word[counter]`
1–2	NA	NA
3–6	0	`'s'`
7	1	`'t'`
4–6	1	`'t'`
7	2	`'h'`
4–6	2	`'h'`
7	3	`'e'`
4–6	3	`'e'`
8–9	3	`'e'`

We iterate through the `while` loop three times, incrementing `counter` to 3. At that point, `word[counter]` is `'e'` and the `if` test is true and we break/exit from the `while` loop. The value of `counter` *at that point* is also the number of consonant characters at the beginning of the word.

The `break` statement also works with a `for` loop. For example, here is the equivalent to the previous example, replacing `while` with `for`:

```
vowels = 'aeiou' #define vowels
word = 'sthenic' #define word
counter = 0      #initialize counter
#go through all letters
for i in range(len(word)):
   #is current letter a vowel?
   if word[i] in vowels:
      break      #if so, exit loop
   #don't forget to update counter
```

```
    counter += 1
#print result
print('The word begins with',
    counter,'consonant letters')
```

control34.py

The logic of the code above is exactly the same as the preceding.

With the `break` statement available to us to exit the `while` loop, we can use `while` tests that are always true, relying on the `break` statement to exit the loop when we want. For example:

```
vowels = 'aeiou' #define vowels
word = 'sthenic' #define word
counter = 0        #initialize counter
while True:        #iterate forever
    #is current letter a vowel?
    if word[counter] in vowels:
        break      #if so, exit
    #remember to increment counter
    counter += 1
#print result
print('The word begins with',
    counter,'consonant letters')
```

control35.py

Note that the `while` test will always be true, so the loop will continue forever unless the `if` test becomes true at some point so that `break` can be executed, exiting the `while` loop. Thus it's important that you be sure `break` is contingent on a test that will be true at some point if the enclosing loop is always true.

The last few examples involve testing letters for whether they are vowel letters, thus the right question is what the code would do if it were fed words with different properties, e.g., a word with no vowel letters. We leave it as an exercise to determine which examples will terminate gracefully and which will either not terminate and loop forever, or terminate with some error.

Slightly different behavior is obtained with the `continue` statement. As with `break`, it is used inside a `for` or `while` loop. What it does is exit the current iteration and go on to the next iteration. Schematically, we have something like this:

```
for ...:
   some statements
   if ...:
      continue
   more statements
```

We have some iterative structure like `for`. We then have zero or more statements. Somewhere in the body we have a `continue` statement, typically in the body of an `if` structure. Following the `if`/`continue`, we have some number of additional statements. What happens is that the `for` licenses some number of iterations. At each iteration, the initial statements apply, then the `if` test occurs. If the `if` test is false, the `continue` statement is not executed and the additional following statements get to apply.

If the `if` test is true on some iteration, then the `continue` statement does apply and the statements following `if`/`continue` do not apply on that iteration. We continue with the next iteration, however, unlike with `break` where the iteration ends. Here's a simple example:

```
vowels = 'aeiou' #define vowels
word = 'sthenic' #define word
counter = 0      #initialize counter
#go through each letter
for i in range(len(word)):
   #is current letter a vowel?
   if word[i] in vowels:
      continue   #if so, skip it
   #increment counter (only for non-vowels!)
   counter += 1
#print result
print('The word has',
   counter,'consonant letters')
```

control36.py

Here we set up some initial variables. We then enter a `for` loop based on the length of the string variable `word`. At each iteration, we test whether the current letter in the string is a vowel. If it is not a vowel, we increment the `counter` variable. If it is a vowel, we go to the next iteration, not incrementing the `counter` variable. In other words, the `counter` variable is incremented only when the current letter is not a vowel. Once we exit the iteration, we print the value of `counter`.

You can also use a `continue` statement inside a `while` structure. The following relatively inefficient code snippet exemplifies this and has the same result as the preceding one.

```
#set up initial variables
vowels = 'aeiou'
word = 'Mississippi'
counter = 0
i = 0
#go through word letter by letter
while i < len(word):
    #is current letter a vowel?
    if word[i] in vowels:
        i += 1          #if so, increment letter count
        continue        #...and exit current loop
    i += 1              #else, increment letter count
    counter += 1        #increment consonant count
#print result
print('The word has',
    counter,'consonant letters')
```

control37.py

3.7 Making Nonsense Items

Let's use the control structures we've learned here to build a program that does something useful. A frequent task for psycholinguists is finding items for experiments, either items that directly exemplify some property we want to test or items that fill out an experiment and can be used to distract subjects from the true goal of the experiment.

Imagine that what we want is CV monosyllables. We could just try to think of all possible monosyllables with a CV shape, but another way to go is to generate these programmatically. If we know what the possible consonants are and what the possible vowels are, we can generate a list of possible CV monosyllables quite simply. Here's a first pass:

```
vowels = 'aiu'          #define vowels
consonants = 'ptk'  #define consonants
for v in vowels:        #for every vowel
    #choose a consonant
```

```
    for c in consonants:
        #now print them together
        print(c,v,sep='')
```

control38.py

This program defines a set of consonants and a set of vowels and then prints out all possible combinations. As such, it's a bit silly. In the case at hand, there are only nine possible combinations, and we could just as easily list those out. This approach becomes more reasonable, though, if the number of consonants and vowels increases:

```
vowels = 'aeiou'         #define more Vs
consonants = 'ptkbdg'    #define more Cs
for v in vowels:         #for every vowel
    #choose a consonant
    for c in consonants:
        #now print them together
        print(c,v,sep='')
```

control39.py

We might also want CVC syllables among our items. This is easy to do as well:

```
vowels = 'aeiou'         #define Vs
consonants = 'ptkbdg'    #define Cs
for v in vowels:         #for every vowel:
    #choose a consonant
    for o in consonants:
        #now choose another consonant
        for c in consonants:
            #print them together
            print(o,v,c,sep='')
```

control40.py

Imagine now we want to exclude cases where the two consonants are the same. That is, we do not want items like *dod* or *bab*. Again, this is straightforward:

```
vowels = 'aeiou'        #define Vs
consonants = 'ptkbdg' #define Cs
#for every vowel, onset, coda
for v in vowels:
   for o in consonants:
      for c in consonants:
         if o == c:   #skip if onset == coda
            continue
         #print combination
         print(o,v,c,sep='')
```

control41.py

Note that this is an excellent example of the difference between `break` and `continue`. If we had used a `break` statement here, we would get the wrong results, exiting the nested `for` structures the first time we ran into identical consonants.

We can extend this to items with more complex structure, e.g., complex onsets, complex codas, and polysyllabic words.

The same kind of logic can be used to construct nonsense sentences. Imagine we want every possible SVO sentence in some (nonsense) language. We have a set of nouns and transitive verbs. We can combine them straightforwardly:

```
nouns = 'bla dor sna'  #every possible N
verbs = 'ha mog ge di' #every possible V
#every possible SVO combo
for s in nouns.split():
   for v in verbs.split():
      for o in nouns.split():
         print(s,v,o)  #print combination
```

control42.py

As in the word-based example above, this is kind of silly when the sets of elements we are combining are so small, but it becomes more useful if we expand those sets.

What do we do if some of our verbs are intransitive? There are a number of ways to deal with this; here's a simple one:

```
nouns = 'bla dor sna'.split()   #Ns
verbs = 'ha mog ge di'.split() #Vs
ivs = 'ha ge'.split()           #intransitives
```

```
#for every N and V
for s in nouns:
   for v in verbs:
      #if V is intransitive
      if v in ivs:
         print(s,v)
      #otherwise it's transitive
      else:
         for o in nouns:
            print(s,v,o)
```

control43.py

Notice the careful placement of the `if` clause above and how it subverts the loop for the object if the verb is intransitive.

Finally, imagine the language has a reflexive pronoun *vi* that replaces the object if it is identical to the subject. We can incorporate this like this:

```
nouns = 'bla dor sna'.split()   #Ns
verbs = 'ha mog ge di'.split()  #Vs
ivs = 'ha ge'.split()           #intransitives
#for every S+V combo:
for s in nouns:
   for v in verbs:
      #if verb is intransitive:
      if v in ivs:
         print(s,v)
      #otherwise (transitives)
      else:
         for o in nouns:
            #if subject and object the same
            if s == o:
               #replace obj with reflexive
               o = 'vi'
            print(s,v,o)
```

control44.py

Notice here how the `if`-test for whether the subject and object are the same is *within* the innermost `for` loop. In this way, we handle every combination of subject and object.

3.8 Summary

In this chapter, we have introduced and exemplified the main control structures of Python: `if`, `for`, and `while`. In conjunction with variable assignment, which we covered in Chapter 2, these constitute the heart and soul of programming in Python. Anything that is computable can be expressed with these bits. The challenge for you in any particular case is to break up the programming task at hand into small pieces that can be expressed with these structures.

To put it directly, understanding variables and control structures is the most important part of learning to program. If you're feeling unsure of these now, there's no shame in going back and rereading these chapters to make sure your foundations are good.

The remainder of this book will provide extensive examples of variables and control structures, so if you're ready to proceed, you'll have lots of opportunities to reinforce these concepts.

3.9 Exercises

3.1 Why does this fail?

```
if 2 + 2 == 5:
print('that shouldn't happen')
    print('or this....')
```

3.2 Augment the recursive prefixation example on page 38 to handle three distinct prefixes. Assume that only identical prefixes can cooccur.

3.3 Augment the recursive prefixation example above to handle three distinct prefixes. Assume that all prefixes can cooccur and any word can have up to three prefixes.

3.4 What's wrong with the following code snippet?

```
count = 0
while count < 3:
    print(count)
    count -= 1
```

3.5 Make a chart that shows how the following would work step by step if we define `word` to be `'cat'`.

```
word = 'alphabet'
count = 0
while count < len(word):
    print(word[count])
    count += 1
```

```
    othercount = 1
    while othercount < count+1:
        print('\t',word[0:othercount])
        othercount += 1
```

3.6 Why does the following code fail?

```
word = 'alphabet'
vowels = 'aeiou'
count = 0
while count < len(word):
    letter = word[count]
    if letter not in vowels:
        print(letter)
        count += 1
```

3.7 The five code examples on pages 44, 45, 46, 47, and 48 involve testing letters for whether they are vowel letters. What happens if the `word` variable has no vowels?

3.8 The program `control41.py` on page 52 uses a `continue` statement. Using a `break` instead would be wrong; why?

3.9 In Section 3.7 of the text, we develop programs to generate nonsense words and nonsense sentences that respect certain restrictions, e.g., avoid identical consonants or replace an identical object with a reflexive. Amplify one of these programs to include another restriction. Make sure to explain what the restriction is!

3.10 **Web**: Snoop around the web to find out how the `switch` control structure works in other programming languages. Python doesn't have it; show how you can simulate its effect with resources Python does have.

4 Input–Output

With variables and control structures, we have the full power of Python available to us. To do anything useful, however, we must be able to run our programs on actual data. For us as linguists, this means words, sentences, texts, sounds, etc.

So far, the only data our programs have been able to manipulate are data that we've coded as part of our programs. For example, the various vowel-counting programs in the previous chapter required that we code the words in which we want to count vowels directly in our program. This is a problem in that we want to write programs like our vowel-counting programs that can count vowels in *any* word.

In this chapter, we learn how to write open-ended programs that can respond to data entered by a user or from a file or set of files. In this chapter, we treat the following ways of inputting data:

Command-line Data can be entered on the command line when the program is invoked.

Standard input Our programs can take input from other programs.

Keyboard input A user can enter data when prompted by the program.

File input–output A program can read data from or write data to files. We will focus on textual data, but this can also include binary data like sound files.

Web Finally, we will show how we can read data in from web pages.

4.1 Command-Line Input

The simplest way to have your programs respond to new data is to enter that data on the command line when you invoke the program. This is quite simple to do. There is a predefined list variable `sys.argv` that contains a list of all command-line arguments given when the program is invoked.

The trick to using `sys.argv` is that it is not, by default, available. To get access to it, you must `import` the `sys` module. As in many other programming languages, Python segregates extra functions and variables in different optional parts of the system. If you want access to these functions or variables, you must make them available to your program with an `import` statement at the beginning of your program. To make `sys.argv` available, we start a program with this statement:

```
import sys
```

If the relevant module is installed on your system, and the `sys` module is a required part of any Python installation, this makes all functions and variables in that module available in the following program. Here's a really simple example:

```
import sys

print(sys.argv)
```

io1.py

We can invoke this program from the command line in different ways and get these results.

```
> python io1.py
['io1.py']
> python io1.py nouns
['io1.py','nouns']
> python io1.py 3
['io1.py','3']
> python io1.py this is a cat
['io1.py','this','is','a','cat']
> python io1.py '3 > 1'
['io1.py','3 > 1']
```

Note first that we are doing this at the terminal (Mac) or DOS (Windows) prompt, *not* in the interactive environment. Second, notice that the first item of the list is always the name of the program. Third, from the quotes, we see that arguments are always interpreted as strings. Finally, certain characters need to be quoted on the command line as they have special interpretations there that we do not want. For example, if we want to enter a command-line argument that contains a greater-than sign, we must quote it as above to prevent the operating system from interpreting it as something else, specifically as redirecting output to a file.

We can use this mechanism to make our vowel-counting program completely flexible. Recall the vowel-counting program `control32.py` on page 45, repeated below:

```
vowels = 'aeiou'          #define vowels
word = 'Winnepesaukee' #set word
#create two counters
counter = 0
vowelcount = 0
#go letter by letter
while counter < len(word):
   #is current letter a vowel?
   if word[counter] in vowels:
      vowelcount += 1
   #keep track of total number of letters
   counter += 1
#when counter is too big, do this:
else:
   print('There are',vowelcount,
      'vowels in this word')
```

control32.py

We can minimally revise this program as follows so that the relevant word comes from the command line:

```
import sys           #make sys.argv available

vowels = 'aeiou'    #define vowels
word = sys.argv[1] #get word from command-line
counter = 0         #proceed as before...
vowelcount = 0
while counter < len(word):
   if word[counter] in vowels:
      vowelcount += 1
   counter += 1
else:
   print('There are',vowelcount,
      'vowels in this word')
```

io2.py

Here we define `word` as `sys.argv[1]`, the second item in the list of command-line arguments. Since the first item is the name of the program, this will be the first thing entered after that. We can produce results like the following:

```
> python io2.py hat
There are 1 vowels in this word
> python io2.py happiness
There are 3 vowels in this word
> python io2.py Appalachicola
There are 5 vowels in this word
```

The program will throw an error if it gets no additional command-line arguments. If it gets more than one, it will disregard all but the first. We can tweak the code to accommodate zero or multiple arguments as follows:

```
import sys          #make sys.argv available

vowels = 'aeiou' #define vowels
#iterate over all words in list
for word in sys.argv[1:]:
   counter = 0    #proceed as before
   vowelcount = 0
   while counter < len(word):
      if word[counter] in vowels:
         vowelcount += 1
      counter += 1
   else:
      print('There are',vowelcount,
         'vowels in',word)
```

io3.py

Recall that `[1:]` returns all items in a list except the first. This program loops over every word in `sys.argv` except for the first, doing the same thing as the previous program. (Recall that the first word in `sys.argv` is the name of the program itself.)

This last program is a nice improvement over the preceding one in that it allows us to count the (lowercase) vowel letters in any number of words. It suffers from several problems, however. First, all words must be entered by hand. If you have a large number of words to treat, this can be prohibitive. A second problem is that there is, in fact, an upper bound on the number of words that can be entered like this. For example, under Windows, the maximum number of characters that can be entered on the command line is just over 8000. It would be quite silly to do that much typing, but our point is that there is an upper bound.

There is another option available that looks a lot like command-line input but gets around these limitations: *standard input* (`stdin`). Any program can output material as what's referred to as *standard output* (`stdout`). That material is typically printed to the screen, and all of our programs so far have provided output of this sort. That output can be read or given as input to other programs. In other words, we run one program, which produces some output, and directly feed that output to a second program. As with command-line arguments, the output of one program is available as input to a second program via a variable from the `sys` module: `sys.stdin`. We gain access to it by again importing the `sys` module. Here's a trivial example:

```
import sys

for line in sys.stdin:
   print(line)
```

io4.py

Note that `sys.stdin` is not a string variable. Instead, it is a *stream*, in effect, a tunnel down which data come. To access that data, we can read it line by line using `for` as in this example.

To use this, we must execute some other command-line program and *redirect* its output to `io4.py`. On Mac, Linux, and Windows, we do this by invoking one command and *piping* its output to our program with the `|` "pipe" symbol. The simplest program we can use for this on any of these operating systems is the `echo` command. This command takes a string and prints it to standard output. If we type this with a string argument at the terminal or DOS prompt, it simply prints that argument:

```
> echo hat
hat
>
```

That's not too interesting in its own right, but that program and argument can be piped to our `io4.py` program as follows:

```
> echo hat | python io4.py
hat

>
```

The outputs are slightly different, as the piped version has an extra empty line. This is because the echo command adds a line, as does the Python print() command.

Notice that echo can send a whole string of words through the pipe, but it does so as a single string, in a single line:

```
> echo hat chair table | python io4.py
hat chair table

>
```

We can revise our vowel-counting program so that it can take multiple words from stdin if we can split that string of words into individual words and then operate on each of them independently. This first step can be achieved with the string method split(). We now tweak io3.py to take multiple words as input from stdin:

```
import sys

vowels = 'aeiou'                    #define vowels
#get each line in stdin
for words in sys.stdin:
    for word in words.split(): #break into words
        #do same as before to each
        counter = 0
        vowelcount = 0
        while counter < len(word):
            if word[counter] in vowels:
                vowelcount += 1
            counter += 1
        else:
            print('There are',vowelcount,
                'vowels in',word)
```

io5.py

Let's go through this line by line to understand. First, as in our previous versions, we define the set of vowels. We next read a line of text from sys.stdin and store that line in the string variable words. The echo command gives us only one line of text, so this part will execute only once. We then split that string into individual words with words.split(), storing those words one by one in the variable word. The rest of the code here simply repeats the same logic from the previous examples, executed on each word into which the string is split.

Here we have read from `sys.stdin` a line at a time using the `for` structure. As noted above, however, the `echo` command, as we've invoked it here, will feed only a single line of text into our program. We can confirm this by tweaking the code in `io5.py` to keep track of what line it's operating on. We do this in `io6.py`:

```
import sys

vowels = 'aeiou'      #vowels
line = 1              #line number
#for each line in stdin
for words in sys.stdin:
   #print line number
   print('This is line',line)
   line += 1          #increment line count
   #break line into words
   for word in words.split():
      counter = 0     #continue as before
      vowelcount = 0
      while counter < len(word):
         if word[counter] in vowels:
            vowelcount += 1
         counter += 1
      else:
         print('\tThere are ',vowelcount,
            ' vowels in "',word,'"',sep='')
```

io6.py

There are several changes here. First, we've added an integer variable `line` to keep track of how many lines we're working on and what the current line is. Second, we've changed the final `print()` command so that it puts double quotes around the word it's working on.

If we invoke the program again, feeding it input from `echo`, we can see that only a single line of text (with three words) is processed.

```
echo cat chair table | python io6.py
This is line 1
   There are 1 vowels in "cat"
```

```
There are 2 vowels in "chair"
There are 2 vowels in "table"
```

Notice that this doesn't help us with the shortcomings of the command-line approach. First, all words still have to be entered by hand. Second, there is an upper bound on the number of words we can enter.

There are other programs, however, that we can feed to our programs via `stdin` to avoid these issues. Another program that will send its output to `stdout` is `cat` (or `type` on Windows). This command prints the contents of a file to `stdout` (the screen). If the file is a parochial word processing file like Microsoft Word, the file contents will be largely uninterpretable. On the other hand, if the file is a plain vanilla text file (typically with the file extension `.txt`), then its output via `cat` is intelligible and can be fed usefully to a program like `io6.py`.

Let's first create a simple text file. Using a text editor, create a file called `test.txt` with these contents:

```
this is
a definite test
file
```

If the file is located in the same directory, we can then print its contents to the screen with this command:

```
cat test.txt
```

Under Windows, we would instead type:

```
type test.txt
```

We can feed the contents of this file to `io6.py` with one of the following commands:

```
#mac or linux
cat test.txt | python io6.py
#windows
type test.txt | python io6.py
```

This produces the following output:

```
cat test.txt | python io6.py
This is line 1
```

```
   There are 1 vowels in "this"
   There are 1 vowels in "is"
This is line 2
   There are 1 vowels in "a"
   There are 4 vowels in "definite"
   There are 1 vowels in "test"
This is line 3
   There are 2 vowels in "file"
```

You see here that `stdin` can accommodate multiple lines when the program feeding it data contains multiple lines.

You should also see that, since files in principle can contain any amount of data, this mechanism allows us to feed an unbounded number of words into our vowel-counting program.

Variable assignment and control structures give us the full power of Python; unbounded input like this allows us to apply that power to a computational problem of any size.

4.2 Keyboard Input

Another way to input data is to request it from the user. That is, you can write programs that pause at some point and wait for the user to enter data. The code for this is quite simple: there is a function `input()` that takes a single string argument. When executed, it prints that string argument and returns what the user types in response as a string. Here's an extremely simple example:

```
theInput = input('Type something: ')
print('You typed "',theInput,'"',sep='')
```

io7.py

The `input()` command prints its string argument to the screen. The program then waits for the user to type something. Once the user hits the return key, the program prints that back with double quotes.

Notice that the string that the `input()` command types does not end with a return or final space by default. The program above adds a space explicitly. If we wanted to, we could add a return instead by explicitly putting that in the string typed:

```
theInput = input('Type something:\n')
```

Notice too that whatever the user types is converted to a string. Thus, if the user types `3`, it will be converted to `'3'`. Hence, if you want the user to enter numbers or data types other than strings, you must include code to convert those. Here's a silly example:

```
#collect two numbers
n1 = input('Enter a number: ')
n2 = input('Enter another number: ')
#convert to integers and add
n3 = int(n1) + int(n2)
print('The sum is:',n3)  #return result
```

io8.py

Entering data like this has similar problems to command-line input: while, in principle, one can enter any number of strings, the data have to be typed in by hand.

On the other hand, there is another potentially desirable aspect of entering data from the keyboard: the number and content of each input item can respond to the program's behavior with respect to earlier items. Here's a silly example of this:

```
import random

letters = 'abcdefghijklmnopqrstuvwxyz'

#get random letter
letter = letters[random.randint(0,25)]

while True:                  #loop until correct
   #prompt them to type a letter
   guess = input('Type a lower-case letter: ')
   #check that it's actually a letter
   if guess not in letters:
      print("That's not a lower-case letter.")
      continue
   if guess == letter:   #if they're right
      print("That's right!")
      break
   #give them a hint if they're wrong
   if guess > letter:
```

```
        print("It's earler in the alphabet.")
    else:
        print("It's later in the alphabet.")
```

io9.py

This program is a guessing game for letters of the alphabet. The program randomly selects a letter, and then the user can guess letters. The interest of the program is that it gives the user feedback on whether their guess is before or after the selected letter. Thus the number and content of each keyboard input is dependent on the program's response to earlier inputs.

There's a lot of code here, but the structure is fairly simple. First, we import from the `random` module to have access to a random number generator: `randint()`. This function generates a random integer between the two integer arguments we give it. Here the range is based on the length of the `letters` string, so we can use the output to index into that string, selecting a single random letter. Notice that the random number we generate with `random.randint()` is immediately fed as an index to `letters`.

We then have an infinite `while` loop with a number of `if` tests. First, we prompt the user to enter a letter and then test that letter. First, we test if the user actually entered a letter. If not, we use a `continue` to exit the current iteration and prompt again. We then test if the user's letter matches the selected letter. If so, we let the user know and exit the loop with a `break`. If it's a legal guess and doesn't match, we then tell the user whether their guess precedes or follows the selected letter alphabetically and continue to the next iteration.

There is one context in which `input()` can be awkward. We've seen that the function returns a string, which we can then convert to a number if appropriate. What if we want the user to enter actual Python variables or functions? For example, imagine we have three variables `x`, `y`, and `z` and we want the user to select one so that the contents of the variable can be printed. Here's the *incorrect* code:

```
#not what we want!
x = 'Tom'
y = 'Dick'
z = 'Harry'

result = input('Type x, y, or z: ')

print(result)
```

io10.py

Here, we might get an interaction like this:

```
> 'Type x, y, or z: x
x
```

To get the right result, we must *evaluate* what the user enters as a Python expression. This can be done with the function `eval()`. Here is the revised code:

```
#set up three variables
x = 'Tom'
y = 'Dick'
z = 'Harry'
#collect user input
result = input('Type x, y, or z: ')
#evaluate and print result
print(eval(result))
```

io11.py

Now we get an interaction like this:

```
> 'Type x, y, or z: x
Tom
```

4.3 File Input–Output

The usual way to input or output large amounts of data is from or to files. The basic idea is that your program is written to respond to any amount of data. The file contains data of the appropriate sort, and your program reads in that data and processes it either all at once or chunk by chunk.

Writing to files, in principle, is a dangerous operation. If you are not careful, you can accidentally overwrite important data. Therefore I recommend several things right at the outset:

(i) Do *not* experiment with important files. Create toy files to play with.
(ii) When you do want to start working on your own files, do *not* use those files directly. It's much safer to create copies of these files and work with those.
(iii) Finally, it's safest to create a new directory to learn file operations in. You can create new files there and copy (not move!) other files there.

These safeguards will reduce the chance of some catastrophic loss of data as you learn file operations.

In principle, the files you work with can be of any sort, but it is best to start with simple text files. Let's begin with writing to a file. The basic logic is that you create a *stream* or *pathway* to a file, print to that stream, and then close the stream. Here's a simple example of this:

```
#open file stream
outFile = open('testfile.txt','w')
#write to it
outFile.write('some text!\n')
outFile.write('...and some more text!\n')
outFile.close()    #close file stream
```

io12.py

First, we create a stream called `outFile`. with the `open()` function. The first argument is the name of the file, and the second argument indicates that we are writing to this file. We then write to that stream twice using the stream method `write()`. Notice that we've explicitly added returns (`\n`) at the end of each `write()` command so that each bit is on its own line in the file. Notice too that each successive `write()` call adds to the existing file. Once we are done writing to the file, we close the stream with the `close()` method.

To beat a dead horse, be careful here. The program above creates a file. If you were to give this file the same name as an existing file in the same directory, you would overwrite the existing file, destroying its contents. Again, create a new directory for file operations at this stage. Also, make sure to name your new test files in a way that is least likely to conflict with your existing files.

Let's now look at file input. The system is basically the same. You create a file input stream, read from it, and then close the stream. In the following example, we read from the file we created in the previous example and print the result to the screen.

```
#open file stream
inFile = open('testfile.txt','r')
stuff = inFile.read() #read form it
inFile.close()        #close stream
print(stuff)          #print contents
```

io13.py

Notice that the `read()` method reads in the entire context of the file. If you want to process the contents of the file in chunks, say lines, this is not optimal.

You have two choices here. One possibility is to break the text into lines after you've read them all in as above. The following program shows how to do this:

```
#open file
inFile = open('testfile.txt','r')
stuff = inFile.read()       #read file contents
inFile.close()              #close file
lines = stuff.split('\n') #split into lines
#print lines and lengths
for line in lines:
   print(len(line),': ',line,sep='')
```

io14.py

In this program we read the entire contents of the file in with the `read()` method. We then use the string method `split()` to break the file contents into lines. We then go through those lines one by one, calculating their length and printing the length and line.

The other possibility is to read lines from the stream and process them one by one.

```
#open file
inFile = open('testfile.txt','r')
#read from stream line by line
for line in inFile:
   #print length of line and the line
   print(len(line),': ',line,sep='',end='')
inFile.close()   #close file stream
```

io15.py

This second program produces similar output. For large files, this second approach can be more efficient since you need not hold the entire file in memory at one time.

So far, we've just looked at text files, but Python can handle specialized or proprietary file formats as well. We give two examples here: wave files and Microsoft Excel files.

Typically, processing files like these requires access to specialized modules that may not be part of your basic Python installation. These can be added in Mac or Linux in two ways, either by general software management programs like MacPorts or Homebrew (for Mac) or by the Python-specific `pip` program. For Windows, `pip` is the normal route. If you're using the Anaconda version of Python, then some modules are available with the `conda` program. In the following examples, we will make use of two such extra modules.

For example, one module we will use below is `openpyxl`, which can be used to read Excel files. To test if the module (or any other) is already on your system, use the `pip` command at the system prompt to list all installed modules. (If you're using the Anaconda Python distribution, replace `pip` with `conda` below.)

```
> pip list
```

You should see a list of all modules currently available on your system.

If this module is not already on your system, you can see if it is available to be installed by typing the following at the prompt:

```
> pip search openpyxl
```

If it is not already installed and you want to install it, you would simply type:

```
> pip install openpyxl
```

Note that if your Python installation is systemwide, you may need to do this last step with admin privileges. For example, on a Mac, you would need to type:

```
> sudo pip install openpyxl
```

You would be prompted for your password. We use this particular module below.

One common file format for linguists is sound files in `.wav` format. These files represent a waveform as a sequence of numbers indicating sound pressure changes over time. In addition, the file contains various sorts of metadata, e.g., sample rate, whether the sound is recorded in stereo, and so on.

We can read in a wave file with a function from the `scipy` library, a specialized module for efficient math functions. There are a variety of things we can do with the data, but the simplest here is just to plot it with a function from `matplotlib`, a specialized module for plotting.

```
#import from scipy and matplotlib
import scipy.io.wavfile,matplotlib.pyplot
#read sample rate and wave vector from file
x,y = scipy.io.wavfile.read('mha.wav')
vdur = len(y)/x                    #calculate duration
#print duration
```

```
print('Duration of wave:',vdur)
matplotlib.pyplot.plot(y)  #make plot
matplotlib.pyplot.show()   #show plot
```

io16.py

This program reads in a wave file using a function from the `scipy` module, which returns the sample rate (in samples per second) and a vector of numbers. Note that if we want to assign these to two different variables, we simply put both to the left of the assignment operator =. The example uses a wave file, `mha.wav` (from among the files on the book website), which is just a recording of the author pronouncing the vowel [a]. We calculate the duration of the wave by dividing the number of samples in the wave by the sample rate. Finally, we use several functions from the `matplotlib` module to create and show the waveform.

We can read in data from other filetypes as well. The following example shows how to use the `openpyxl` package to read in and examine an Excel spreadsheet.

```
import openpyxl  #to handle xls/xlsx files

#read in data
wb = openpyxl.load_workbook('test.xlsx',
   read_only=True)

#get the names of 'sheets'
print(wb.get_sheet_names())
#get the first sheet
sheet = wb['Sheet1']
#print contents of cell B2 on sheet1
print(sheet['B2'].value)
r = 0            #keep track of rows
#go through all rows
for c in sheet.rows:
   print(r)        #print the row number
   #print cells in each row
   for i in range(len(c)):
      print('\t',c[i].value)
   r += 1
```

io17.py

Here we first load in a simple spreadsheet we've created, `test.xlsx`, with a function from the `openpyxl` module. Excel spreadsheets contain multiple pages or *sheets* with grids of cells that can be filled with data of different types. We first extract the names of the sheets in this spreadsheet. We then extract the (only) one named `Sheet1` and store that in `sheet`. We can access individual cells by name or we can iterate through all rows, printing out the contents of all cells in those rows.

4.4 Alice in Wonderland

In this section, we write a larger program to do lexical statistics on *Alice's Adventures in Wonderland* by Lewis Carroll.[1]

Our first step is to make sure we can read in the file. Let's just do that and count the lines in the file. Here's one way to go about it:

```
count = 0                      #counter for lines
f = open('alice.txt','r')      #open the file
for line in f:                 #read line by line
    count += 1
f.close()                      #close file
print('lines:',count)          #print line count
```

io18.py

Let's now save all the lines in a list:

```
count = 0          #counter for lines
lines = []         #list for line contents
#open file
f = open('alice.txt','r')
for line in f: #read it line by line
    count += 1   #add 1 to counter
    #add current line to list
    lines.append(line)
f.close()          #close the file
#print number of lines read
print('lines:',count)
#print number of lines saved
print('saved lines:',len(lines))
```

io19.py

[1] The full text is available from Project Gutenberg (`http://gutenberg.org`) and is included on the book website.

In this latter example, we have created an empty list and then added the lines one by one to the end of that list. At the end of the program we print out the number of lines read and the number of lines in the list. If we've done everything correctly, those two numbers should be the same. This is good programming practice generally. Print out the values of things as you proceed so that you can make sure the program is behaving as you intend.

Let's now print out the first few lines:

```
lines = []          #list to save lines
#open file
f = open('alice.txt','r')
for line in f: #read line by line
    #save each line in list
    lines.append(line)
f.close()           #close file
i = 0               #print first 100 lines
while i < 100:
    print(lines[i])
    i += 1
```

io20.py

The result here is not quite right; each line is printed out with an extra line in between. The problem is that lines are read in with their final return character. The `print()` function supplies another return, and each line is then terminated by two return characters. We can get the behavior we want by telling `print()` not to append a return.

```
lines = []          #list to save lines
#open file
f = open('alice.txt','r')
for line in f: #read line by line
    #save lines to list
    lines.append(line)
f.close()           #close file
i = 0               #print first 100 lines
while i < 100:
    #don't add a return to the line!
    print(lines[i],end='')
    i += 1
```

io21.py

Notice now that the lines we're printing out are not part of the *Alice* story, but of a header that Project Gutenberg has added to the beginning of the file. By playing around with the number of lines we print out, we can see that the header is 255 lines long. Our next version of the program removes this header and then prints out the beginning of the story:

```
lines = []          #list for lines
#open file
f = open('alice.txt','r')
for line in f: #read lines one by one
    #add lines to list
    lines.append(line)
f.close()           #close file
#strip off first 255 lines
lines = lines[255:]
i = 0               #print first 50 lines
while i < 50:
    #still don't add a return!
    print(lines[i],end='')
    i += 1
```

io22.py

Let's now do some analysis of the lexical content of the file. As a simple example, let's imagine that we are interested in whether there is a correlation between word length and word frequency. It's generally believed that more frequent words are shorter than less frequent words. To look at this with our text, we must break each line into words and then compute the length of each word. We'll keep track of how many words we see of each length.

The next version of the program breaks each line into words and then stores all the words in a list.

```
words = []          #list of all words
lines = []          #list of all lines

#open file
f = open('alice.txt','r')
for line in f: #save lines one by one
    lines.append(line)
f.close()           #close file
```

```
#remove Gutenberg header
lines = lines[255:]

#go through lines one by one
for line in lines:
   #break each line into words
   wds = line.split()
   #add words to list
   words += wds

i = 0 #print first 100 words
while i < 100:
   print(i,words[i])
   i += 1
```

io23.py

This program does indeed get all the words, but it doesn't strip out irrelevant punctuation. Words are returned with adjacent punctuation like period, question-mark, etc. We need to strip these away before doing our counts if we want an accurate picture of the relationship between word length and frequency.

There are better ways to do this, as we'll see in Chapter 6. Our approach here will be to go through each word character by character, counting only alphabetic characters and nothing else. To make this task easier, we first convert words to lowercase.

To test this idea and make sure it's doing the right thing for us, we'll first write some code that does this for the first 100 words and displays the output. If it works, we then will scale it up for all the words in the file. Here's a program that shows how to do this:

```
words = []      #list of all words
lines = []      #list of all lines

#open file
f = open('alice.txt','r')
for line in f: #save lines one by one
   lines.append(line)
f.close()       #close file

#remove Gutenberg header
lines = lines[255:]
```

```
    #go through lines one by one
    for line in lines:
        wds = line.split()          #break into words
        words += wds                #add to list

    #print first 100 words and letter counts
    i = 0
    while i < 100:
        #store the count for the current word
        count = 0
        #convert the current word to lowercase
        word = words[i].lower()
        #go through word letter by letter
        #if lowercase, add 1 to count
        for l in word:
            if l in "abcdefghijklmnopqrstuvwxyz":
                count += 1
        print(i,words[i],count) #print it all
        i += 1
```

io24.py

If you inspect the output of this program, you'll see that it does get the correct letter count for the first 100 words. Given that that part is doing the right thing, we can now scale up to doing this for all words and saving the results. What we want to know is how many words there are of each length. To do this, we construct a dictionary that we'll use to store the number of words we've seen for each word length. If, for example, we were to call this dictionary `wordlengths`, we would have the number of words that are two letters long in `wordlengths[2]`, etc.

The following program implements this idea:

```
words = []          #list of all words
lines = []          #list of all lines
wordlengths = {} #dictionary of word lengths

#open file
f = open('alice.txt','r')
for line in f:      #save lines one by one
    lines.append(line)
f.close()           #close file
```

```
#remove Gutenberg header
lines = lines[255:]

#go through the lines one by one
for line in lines:
   #break each line into words
   wds = line.split()
   #add the words to the list
   words += wds

for wd in words:
   count = 0       #count for current word
   #convert current word to lowercase
   word = wd.lower()
   #go through word letter by letter
   #if lowercase, add 1 to count
   for l in word:
      if l in "abcdefghijklmnopqrstuvwxyz":
         count += 1
   #check if we've seen this length before
   if count in wordlengths:
      #if so add 1
      wordlengths[count] += 1
   else:
      #if not, set to 1
      wordlengths[count] = 1

#print out counts for each word length
for c in wordlengths:
   print(c,wordlengths[c])
```

io25.py

Finally, let's have the program save the results in a file:

```
words = []         #list of all words
lines = []         #list of all lines
wordlengths = {} #dictionary of word lengths

#open file
f = open('alice.txt','r')
#save lines one by one
```

```
for line in f:
   lines.append(line)
f.close()          #close the file

#remove Gutenberg header
lines = lines[255:]

#go through lines one by one
for line in lines:
   #break each line into words
   wds = line.split()
   #add words to the list
   words += wds

for wd in words:
   count = 0      #count for current word
   #convert current word to lowercase
   word = wd.lower()
   #go through word letter by letter
   #if lowercase, add 1 to count
   for l in word:
      if l in "abcdefghijklmnopqrstuvwxyz":
         count += 1
   #check if we've seen this length already
   if count in wordlengths:
      #if so add 1
      wordlengths[count] += 1
   else:
      #if not, set to 1
      wordlengths[count] = 1

#open output file
g = open('res26.txt','w')
#print out counts for each word length
for c in wordlengths:
   clen = str(wordlengths[c])
   res = str(c) + ': ' + clen + '\n'
   g.write(res)
g.close()          #close output file
```

io26.py

This program does quite a bit, and we have built it up step by step from the building blocks you have learned thus far. Creating the program in this way does two things that we need to keep sight of.

First, it shows us the contribution of each program element separately. This way, we can examine and understand what each is doing.

Second, and more importantly, the stepwise construction of this program is a model for how *you* should write your own programs. You should build them up step by step, checking at each point that your program is doing what you think it should be doing. You check this by printing the value of variables at each point and checking that they are what you want. If they are, you strip out those print statements and go on to the next step, printing out the new variables of interest.

This style of building programs is not just for beginners. I continue to use it myself in my own programming practice, and I recommend you get comfortable with it now and make it part of your programming habits.

4.5 Summary

In this chapter, we've talked about how to input data into programs. The basic idea is to write programs that can respond to novel data and/or any amount of data. Any number of data items can then be fed into the program.

We considered a number of input forms. First, the number of forms can be entered on the command line when your program is invoked. Second, programs can be fed into your program via standard input (`stdin`). Third, the user can be prompted for input. Finally, data for your program can come from files of various sorts. We also saw that you can output your data to files.

4.6 Exercises

4.1 Tweak the `io3.py` program to accommodate uppercase vowels.

4.2 Tweak the `io3.py` program to count both vowels and consonants. Make sure it can handle both uppercase and lowercase letters.

4.3 Tweak the `io3.py` program to count letters from a list the user supplies. Thus the first command-line argument is a sequence of letters. These are the letters the program will count. The remaining words on the command line are those that the counts are performed on.

4.4 What happens if we feed data into `io6.py` and then *again* into `io6.py`?

```
... | python io6.py | python io6.py
```

4.5 The programs `io14.py` and `io15.py` produce slightly different output. Why?

4.6 Write a program that takes input from `stdin` and converts what it gets to lowercase. Show how this program can be used transitively, in between two other programs, i.e., `prog1 | prog2 | prog3`.

4.7 Write a program that takes a simple mathematical expression on the command line like `'7 / 45'`, parses it correctly, and prints the result.

4.8 Write a program that goes through the Alice text and prints out all multiple *wh-* questions. We define these as questions that begin with a *wh-* word and contain at least one more *wh-* word in the question. (If you already know what they are, you may not use regular expressions.)

4.9 Write a program that:

(a) Reads in the *Alice* file;
(b) strips off the header;
(c) converts everything to lowercase;
(d) converts all punctuation to spaces;
(e) splits the text into words;
(f) counts up how frequent all words are;
(g) prints out the 10 most frequent words and their counts.

4.10 **Web**: The `write()` function will also let you append to an existing file. Snoop around on the web for how to do this and write a brief program that exemplifies.

5 Subroutines and Modules

We can now write fairly large programs. As they get larger and longer, it becomes harder to understand them, harder to keep track of what they're doing, and harder to make changes when changes are necessary.

There are several ways you'll see this effect in your actual code. First, code will become repetitive; you'll find yourself repeating whole blocks of code in your programs. This is always a bad idea. The problem is that if you ever need to change one of these blocks, you'll typically need to change all of them, and it's quite easy to either forget to do that, or to be inconsistent about it.

Another way you'll see this in a language like Python is that your indentation for code blocks will become confusing. You'll be working on some line of code at the end of a block that's tabbed/spaced in an unspecified number of times and not know how many tabs or spaces you need for the next line.

These issues make us want to *factor* our code, break it into more efficient and conceptually more reasonable chunks. In this chapter, we introduce two ways to do this: *functions* and *modules*. We've already been using functions that are provided directly by Python along with functions available in modules we've imported. In this chapter, we show you how to write your own functions.

We've already seen how to use modules, either those part of your default Python installation or publicly available modules you add to your installation. In this chapter, we show you how to write your own modules.

The chapter is organized as follows:

Simple functions We give the basic syntax for functions and show how they can be used to simplify our code.

Functions that return values We've seen functions that can "do" things, like the `print()` function, but there are also functions that give us things, like the `len()` function. In this section, we show how to write functions that return values.

Functions that take arguments Functions may or may not take arguments, and here we show how to add required or optional arguments to your functions.

Recursive and lambda functions In this section, we go on to recursive and anonymous (lambda) functions. These topics are a bit advanced, but if you've had a fair amount of syntax or semantics, they are quite straightforward.

Modules Next, we show how to package a set of functions and variables into separate modules that can be used by multiple programs.

Docstrings and comments Finally, we enrich our understanding of comments and include discussion of *docstrings*, specialized comments that can be displayed via Python's `help()` function.

5.1 Simple Functions

Functions are defined with the `def` keyword, followed by a function name of your choosing, parentheses, and a colon. These are followed by a block of statements.

```
def myfunction():
    print('This is a function')
    print("That's all it does")
```

func1.py

We can invoke this function just as we invoke any other function. The invocation must *follow* the function definition (but need not follow it immediately).

```
def myfunction(): #function definition
    print('This is a function')
    print("That's all it does")

myfunction()        #invoking the function
```

func2.py

Functions allow us to simplify repetitive code. Consider the following simple program:

```
#print a famous sentence over two lines
print('Colorless green ideas...')
print('...sleep furiously')
#get the user to enter a number
num = input('Enter a number: ')
```

```
#print the sentence again if it's < 5
if int(num) < 5:
   print('Colorless green ideas...')
   print('...sleep furiously')
else:
   print('Your number was big enough')
```

func3.py

Notice how the first and second lines are repeated in the `if` block. We can avoid this repetition by defining our own function:

```
#a function to print that sentence
def myfunc():
   print('Colorless green ideas...')
   print('...sleep furiously')

myfunc()     #invoke the function
#collect the number
num = input('Enter a number: ')
#check if the number is < 5
if int(num) < 5:
   myfunc() #print sentence again if so
else:
   print('Your number was big enough')
```

func4.py

Here we factor out the repeated two-line section as a separate function. We then call that function twice in the following code.

In this case, we have the same number of lines overall, but you can see that as the number of repetitions increases or as the size of factored-out code increases, we achieve savings in terms of the number of lines in the overall program.

Savings in terms of number of lines is not the point, however. First, the new code in `func4.py` makes clear that those repeated parts are the same. The new code is also easier to maintain in the sense that if we want to change one of the repeated lines, we can do so *once* and the change is applied in *both* applications of the function.

Functions can be more complex. Here's a slightly longer bit of code that takes input from the user.

```
def myfunc():            #a function
   #user supplies a word
   word = input('Word: ')
   #print that word
   print('This is your word:',word)
   if len(word) > 5:    #check if > 5
      print('Your word was long.')
   else:
      print('Your word was short.')

myfunc()                 #invoke function
```

func5.py

Here the function reads user input, prints it, and then does different things depending on the length of the input.

Notice how the function code creates a variable `word`. It's important to note that that variable is available *only inside the function*. If we try to refer to a variable created and assigned a value inside some function outside that function, we get an error. Here's an example of that:

```
#this doesn't work!

def myfunc():
   word = input('Word: ')
   print('This is your word:',word)

myfunc()
if len(word) < 5:
   print("Your word wasn't long enough")
```

func6.py

On the other hand, variables outside a function are available inside a function. Here's an example:

```
word = input('Word: ') #user supplies word

#function refers to previous value!
def myfunc():
   print('This is your word:',word)
```

```
myfunc()                    #invoke function
#check if word is less than 5 letters
if len(word) < 5:
   print("Your word wasn't long enough")
```

func7.py

Here the value of `word` is set outside the function. When the function is called, it has access to that value. It is generally a *bad* idea to refer to external variables like this in a function, since it can lead to errors of various sorts. A better and safer approach is discussed in Section 5.3.

5.2 Functions That Return Values

So far, our functions have taken no arguments and have returned no values. In this section, we show how to write functions that return a value. The syntax is simple: the returned value of a function follows the keyword `return`. Here's an example:

```
def myfunc():                #function definition
   print("This prints.")     #prints this
   return 6                  #return the value 6
   #gratuitous print command
   print("This doesn't print!")

#invoke function, assign value to x
x = myfunc()
#print value of x
print("Here's the function output:",x)
```

func8.py

Notice that the string `"This prints."` is *not* the value returned by `myfunc()`. Rather it is simply something that is printed when the function runs. The value returned by the function is the one that appears after the `return` statement: `6`. Notice too how the first `print()` statement executes, but the second does not. The `return` statement exits the function, so subsequent statements cannot run.

This does not mean that `return` must always be last in the function. We can, for example, embed `return` in an `if` structure.

```
def sillyfunc():      #function definition
   #user supplies a word
   wd = input('Type a word: ')
   if len(wd) > 4:   #check length of word
      #return length and exit function
      return len(wd)
   else:             #otherwise...
      print('The word is too short!')

res = sillyfunc()    #save value of function
#print value of variable
print('The result: ',res)
```

func9.py

Note that this function either returns a value or prints a value. If you use it in a context where you expect it to return a value, with = for example, you end up with the non-value `None` if it does not, in fact, return a value.

Functions can also return more than one value. This is as simple as putting several values – separated by commas – after the `return`. Here's a simple example:

```
def myfunc():         #function definition
   #collect two strings
   x = input('First string: ')
   y = input('Second string: ')
   z = x + ' ' + y #concatenate strings
   #return all three
   return len(x),len(y),z

#invoke function saving all
a,b,c = myfunc()     #3 return values
print('a =',a)       #print the 3 values
print('b =',b)
print('c =',c)
```

func10.py

Here the function prompts for two strings and then returns three values: the length of the first string, the length of the second string, and the concatenation of the two strings. Note, incidentally, that the returned values need not be of the same type. Here, the first two returned values are integers and the third is a string.

5.3 Functions That Take Arguments

Functions can take arguments as well. This is superficially simple to do, but can get tricky. Basically, for a function to take an argument, you put *new* variable names in the parentheses in the function definition. Here's a simple example:

```
#function that takes 2 arguments
def myfunc(a,b):
    #return the concatenation
    #OR addition of those values
    return a + b

#invoke the function with numbers
print(myfunc(3,10))
#invoke the function with strings
print(myfunc('strings ','too'))
```

func11.py

In this example, we give our function two arguments, a and b. The function then applies the + operator to those. If we invoke the function with integer arguments, the result is the addition of the arguments. If we invoke the function with string arguments, we get the concatenation of those arguments. Thus, Python generally does not restrict the type of the arguments a function can take.

Ideally, argument names should be novel, not to be confused with other variables. Let's consider this a bit more closely, however. Imagine we have a function defined in a context like the following:

```
x = ...

def afunction(y):
    ...
    z = ...
    ...
    return z

w = afunction(x)
```

In this schematic example, x and w are global variables – variables defined outside of any function and available throughout this file and in any functions defined in this file. The variable z is defined within the function afunction and is *not* available outside that function. This, however, does not prevent the *value* of z from being returned by the function.

Finally, there is the variable y. This variable is available only within the function afunction and is *not* available outside of it. The value of y is taken from whatever element is given as an argument to afunction when it is invoked.

Here things get a little tricky, and it's important to remember our discussion of mutability in Section 2.4. Consider first the following program:

```
x = 'a value'

def anotherfunc(a):
    a = 'another value!'
    return a

print(x)
print(anotherfunc(x))
print(x)
```

func12.py

Here we assign a string to the variable x. We then define a function that takes an argument. We reassign a value to that argument *inside* the function and then return it. After the function definition we print the variable, print the output of the function, and then print the variable again. This produces the following output:

```
> python func12.py
a value
another value!
a value
```

Recall that strings are immutable. Hence, when we change the value of a in the function, we're not really changing the value of a at all; instead, we're assigning a new value to a and leaving the old one available for garbage collection. In this case, however, the old value is still attached to x, so it isn't collected as garbage. Hence, when we print the value of x after the function applies, it retains its original value.

Compare this with the following program:

```
x = [4,5,6]

def anotherfunc(a):
    a.append(7)
    return a

print(x)
print(anotherfunc(x))
print(x)
```

func13.py

Here, we assign a list to x. We then define a function that takes an argument and we append 7 to that argument and return the result. After the function definition, we then print the value of x, print the output of the function, and print x again. This gives us this output:

```
> python func13.py
[4, 5, 6]
[4, 5, 6, 7]
[4, 5, 6, 7]
```

Notice here that the value of x changes after the function applies! This is because lists are mutable. Hence, when the function applies, it forces a to refer to the same list as x. We then change that list so that when we print x after the function applies, the x value has changed as well.

The difference between these two programs corresponds to mutability. Mutable objects all work as in the second program; immutable objects work as in the first one. (If this is murky, it might be a good idea to go back and review Section 2.4.)

We've already seen functions with more than one argument. There is more than one way to invoke these. Consider this example:

```
#function definition
def thefunction(x,y):
    return x + ' ' + y

#invoke the function 3 ways
print(thefunction('one','way'))
```

```
print(thefunction(x='another',y='way'))
print(thefunction(y='way',x='yet another'))
```

func14.py

Here we define a function with two arguments and invoke it in three different ways. One way is to just provide two arguments; these are interpreted as the two arguments in the function in the same order. A second way is to name the variables. The names tell us which argument gets associated with which variable. Note that if we take this option, we can give the names in any order, as above.

The = can be used in the function declaration as well to give default values to the function arguments. Here's an example:

```
#function with default for 2nd arg
def f(x,y='oops'):
   return x + ' ' + y

print(f('hat'))        #invoked 3 ways
print(f(x='chair'))
print(f('hat','chair'))
```

func15.py

Here we define a function that takes two arguments and concatenates them. The second argument has the default value `'oops'`. We then invoke the function three times with this output:

```
> python func15.py
hat oops
chair oops
hat chair
```

In the first invocation, we give the function one argument, which is taken as the value for `x`. The argument `y` takes its default value. The second invocation has the same effect. In the third invocation, we give two arguments, so `x` is associated with the first and `y` with the second.

Note that if you mix arguments with and without default values in your function declaration, all the arguments with default values must follow the arguments without.

You can also write functions with an unspecified number of arguments. If you want an unspecified number of unnamed arguments, you put the variable name in the parentheses in your function declaration with a preceding asterisk. This variable can then be used as a list in the function body. If you want an unspecified number of named arguments, you put the variable name in the parentheses in your function declaration with two preceding asterisks. This variable can be used as a dictionary in the function body. If you use both, the list variable must precede the dictionary variable. Here is a simple example with both:

```
#function with unspecified
#number of unnamed and named arguments
def func(*args,**kwargs):
   for a in args:   #print unnamed args
      print(a)
   for k in kwargs: #print named args
      print(k,'\t',kwargs[k])

#invoked with unnamed FOLLOWED by
#named arguments
func(3,6,8,hat='wow',chair=3.5)
```

func16.py

This function simply prints out all variables given as unnamed variables and then prints out all variables given as named variables. The latter are printed with their names.

5.4 Recursive and Lambda Functions

This section contains some tricky material. This section is *not* critical to getting started programming, so it can be safely skipped on a first read. If you've had some formal semantics on the other hand, or background in functional programming, then enjoy.

A useful feature of Python is that functions can be manipulated just like other data types. Thus, for example, it's possible to give functions as arguments to other functions. Here's a simple example:

```
#function with 2 args:
#   a function f
#   and something else x
def func(f,x):
   return f(x)

print(func(len,'hat'))  #invoking it
```

func17.py

This program defines a function that takes two arguments: a function and something else. It then applies the first argument, the function, to the second argument and returns the result. It then applies this function to the arguments `len` and `'hat'`. Notice here that the argument names are entirely arbitrary. Here we choose `f` and `x`, but we could have just as easily chosen `harry` and `ernestine`. Using `f` for the function variable just adds a bit of clarity to what we want this function to do.

Functions can also return other functions. Here is an example:

```
#function that returns a function
def func(x):
   if x == 'L':
      return len
   else:
      return type

#invoking the function returns a
#function which we apply to
#'chair'. This may look confusing....
print(func('L')('chair'))
```

func18.py

This program defines a function that takes a single argument. If that argument is the character `'L'`, then the function returns the function `len()`; otherwise, it returns the function `type()`. We then apply this function to the argument `'L'` and then apply the result of that application to the string `'chair'`.

Functions can, in fact, manipulate *themselves*. A function definition that does this is said to be a *recursive function*. A typical example of this in many programming languages is a function for calculating factorials.[1] Recall that factorials are defined like this:

$$\begin{aligned} 1! &= 1 &&= 1 \\ 2! &= 2 \times 1 &&= 2 \\ 3! &= 3 \times 2 \times 1 &&= 6 \\ 4! &= 4 \times 3 \times 2 \times 1 &&= 24 \\ 5! &= 5 \times 4 \times 3 \times 2 \times 1 &&= 120 \end{aligned}$$

Basically, we define 1! as 1. We then define any higher number n as $n \times (n-1)!$. Here is a function that does this:

[1] This may seem like an excessively mathematical example, given our focus on language-related programming, but in fact factorials do show up in phonological theory. In Optimality Theory, if you have n constraints, then there are $n!$ possible rankings.

```
def fac(n):          #function definition
   if n == 1:        #base case of recursion
      return 1
   else:             #recursive clause
      #invokes the function ITSELF
      return (n * fac(n-1))

#invoked with base case
print('1! =',fac(1))
#invoked with recursive case
print('5! =',fac(5))
```

func19.py

Here we define a recursive function `fac()`. If its argument is `1`, then we return `1`. If its argument is anything else (presumably a higher integer), then we return that value times the result of applying `fac()` to the next lower integer. Let's walk through how this works for the invocation `fac(3)` in the table below.

First, we invoke the function with the call `fac(3)` in line a. Since the argument is not `1`, this is converted to line b. Line b includes the call to `fac(2)`, which is converted to 2 × `fac(1)` in line c. Finally, `fac(1)` is converted to 1 in line d.

a.	`fac(3)`	=	
b.	3 × `fac(2)`	=	
c.	3 × 2 × `fac(1)`	=	
d.	3 × 2 × 1	=	6

Finally, Python allows you to create unnamed functions on the fly and manipulate or apply them. Here's a simple example:

```
print((lambda x: x + x)('hat'))
```

func20.py

Here we define an anonymous function that concatenates a string with itself. We then apply that anonymous function to the string `'hat'`. The syntax is just like a regular function except that the keyword `def` and the function name are replaced with the keyword `lambda`. The resulting expression can then be used just like any other function by putting parentheses with any arguments after it.

Here is another more interesting example:

```
def makeAddN(n):     #function definition
   #returns new function
   return lambda x: x + n

#invoke twice, making 2 new functions
add2 = makeAddN(2)
add6 = makeAddN(6)
#apply those two new functions
print(add2(17))
print(add6(17))
```

func21.py

In this case, we have defined a function that creates and returns new functions. The function-creating function takes a single integer argument and then returns a new function that adds that integer to its argument. We then create two new functions and apply them to 17.

5.5 Modules

We've seen that Python has various objects, functions, and methods that are available from the get-go and that it has other objects, functions, and methods that are available only when an appropriate `import` statement has occurred. For example, we saw that the list of command-line arguments `sys.argv` is available only when we have included `import sys` at the beginning of the program. Similarly, we saw that the `randint()` function is available only when we have imported the `random` module.

This may seem like a peculiarity, but in fact is deliberate and efficient. The basic idea is that some programming elements are quite frequently needed while others are needed only in specific situations. The module system separates things accordingly. General functions and objects are available in all cases, but you make more specific functions and objects available only when you are writing specific sorts of programs.

The alternative would have all programming elements available all the time. This would amount to thousands of functions being available at once. It would mean you'd have to be careful not to overwrite them with your own variables and function definitions. In addition, we would have to have sufficient names to distinguish everything from each other, which would make for rather long names as well.

In this section, we first show how to find out what modules are available to you and how to get help on any of them. We then show various ways to import modules. In the next section, we show how to write your own modules.

To find out what modules are installed on your system, go to the interactive system and type:

```
>>> help('modules')
...
```

This will generate a list of every module installed by default and modules you have added. In addition, it will list the names of all Python programs in your current directory. To find out more about any one of these, you first import it and then use the help system. For example:

```
>>> import re
>>> help(re)
...
```

When you import a module, the objects and functions of that module are available with the name of the module prefixed on the left separated by a period. Thus when we make use of `sys.argv` in the `sys` module, this is because the object `argv` is defined within the `sys` module. We saw this in the `io1.py` program, repeated below.

```
import sys

print(sys.argv)
```

io1.py

Anything within the `sys` module can be accessed like this. Alternatively, we can import only specific elements of a module. When we do this, we need not specify the full module name in naming the object or function. The following example shows this:

```
from sys import argv

print(argv)
```

func22.py

Here we have access to `argv` from the `sys` module, but we need not give the full name when we use it. In addition, nothing else from that module is available since we have not explicitly imported anything else. Notice that if we do this sort of restricted import, the full name will not work:

```
from sys import argv  #doesn't work!

print(sys.argv)
```

func23.py

In fact, we can import everything from a module with this syntax by replacing `argv` with `*`. This is generally not a good idea as it makes everything in the module available and can create unintended name conflicts.

We have one more option with `import`. We can create an alias for the module in the `import` statement. We can do this as follows:

```
import sys as s

print(s.argv)
```

func24.py

This allows us to use a different module prefix for any function or object we might import. Altogether, the different import options allow us to keep our program *name space* as uncluttered as possible, both by restricting what elements are available and by letting us control how specific the names of those elements are.

5.6 Writing Your Own Modules

Writing your own modules that can be imported by other programs is as simple as writing programs. Let's create a module with a single function and string variable in it:

```
#our own module

myVar = 'hats and lemons'  #variable

def myFunc(s):             #function
   return len(s)
```

func25.py

We can call import and use this variable and this function in all the ways we saw above. First, we can import and use full names:

```
import func25          #import function

#invoke variable with full name
print(func25.myVar)
#invoke both with full names
print(func25.myFunc(func25.myVar))
```

func26.py

We can also import specific elements or all elements. These can then be used without the module prefix:

```
#invoke everything from the module
from func25 import *

#invoke variable without prefix
print(myVar)
#invoke both without prefixes
print(myFunc(myVar))
```

func27.py

Finally, we can, as above, import the module with a different name:

```
#import with abbreviated prefix
import func25 as f

#invoke function with f prefix
print(f.myVar)
#invoke both with f prefix
print(f.myFunc(f.myVar))
```

func28.py

Notice that if we run `func25.py` on its own, we get nothing, as that file defines a function and a string variable but does nothing with them. We can, in fact, write modules that can be imported *or* be run on their own. When imported, they provide functions and variables that other programs can use.

When they're run on their own, they can use those functions and variables themselves.

To do this, we take advantage of the `__name__` variable. When a program is loaded in directly, rather than imported from another program, the `__name__` variable is automatically set to `'__main__'`. We can then use this to allow a module to behave differently when it is invoked directly versus when it is imported by another program. Here's a simple example:

```
#module that can run on its own

myVar = 'hats and lemons'   #variable

def myFunc(s):              #function
   return len(s)

#if this is loaded on its own...
if __name__ == '__main__':
   print(myFunc(myVar))     #do this
```

func29.py

This program will print out the length of `myVar` when it is invoked on its own. On the other hand, when imported by another program, it will behave just like `func25.py`.

At this point we need to revisit the notion of commenting our code. So far, we have used comments in programs as notes to ourselves, or perhaps as notes to other programmers who might inherit our code. When we start writing modules, the need for comments changes slightly. We now want comments that will enable other programmers to use the functions and variables our module provides.

The standard way to document our functions for other programmers is to use *docstrings*. When you are in the interactive environment, you can use the `help()` function to find out more about any function. Docstrings are what allow you to do that. Basically, a docstring is a triple-quoted string that occurs within a function right after the `def` line. For example, imagine we write a module like the following:

```
def myLen(s):
   '''This computes the length of a string.

   s -- the string
```

```
    '''
    return len(s)
```

func30.py

The triple-quoted string includes two bits of information: an intuitive statement of what the function does and an explanation of what the argument is. We can now get help on this function, either in the interactive environment or in a separate program that imports `func30.py`:

```
import func30

help(func30.myLen)
```

func31.py

A docstring is what you use if you want to make the functions of your modules usable by others. A comment is what you use for you or others to understand or alter your code. Thus the functions in your modules should make use of both, but in different ways. We discuss this further in the next section.

We've talked about writing modules that make functions and variables available for other programs. We've also discussed how to document the functions your modules provide. We also want to consider the case where your module includes functions or variables that you *don't* want to make available.

This may seem odd, but in fact is perfectly reasonable. Imagine you've written a module that is meant to provide some specific functionality, for example, a function to read in a file and return a list of all words with an even number of letters and how frequent each of those words is. Your module would then optimally make available a function with a name like `evenCount()` that takes a filename as an argument and then returns, for example, a Python dictionary where each key is a word with an even number of letters and each value is the frequency of that word.

To make that optimal function work, however, there may be a number of other "helper" functions that you've written. These would be in the module and called by the `evenCount()` function, but you don't really intend other programmers to have direct access to them. Such private helper functions should always have names that begin with _.

Here's a simple example. In this program we define a function `myF()` that returns the number of words in a string minus one. We've written this somewhat clumsily to take advantage of a helper function, `_mySplit()`. This latter function splits a string into words and returns all the words except the first one. We've written this latter function to be private.

```
def myF(s):          #this uses _mySplit()
    '''This calculates the number of
    words in a string minus one.

    s -- the string
    '''
    wds = _mySplit(s)
    return len(wds)

def _mySplit(s):     #this is private!
    '''This returns all the words in
    a string except the first.

    This docstring is pointless!
    '''
    ws = s.split()
    return ws[1:]
```

func32.py

When a function is private, it is not available from *certain* calls to `import`. For example, this code produces an error:

```
from func32 import *  #doesn't work!

print(_mySplit("This doesn't work"))
```

func33.py

On the other hand, this code works:

```
import func32

print(func32._mySplit("Oh, this does work"))
```

func34.py

As does this:

```
from func32 import _mySplit

print(_mySplit("Oh, this works too"))
```

func35.py

Effectively, there is a limited amount of privacy granted by using the underscore prefix. As we've seen, some ways of importing will allow one to reach into a module and make use of something with an underscore prefix. Aside from how the language works, Python programmers operate on the general assumption that functions and variables that begin with underscore are *intended* to be private and *intended* not to be imported. If you use underscore in this way, it will help other programmers understand the logic and intent of your code.

5.7 Docstrings and Modules

Modules are documented with *docstrings* as well: triple-quoted strings that appear at the beginning of a module or as the first line of a function.[2] Following is a simple example of both:

```
'''This is a test module.

Author:
   Mike Hammond
Version:
   1 (11/18)
'''

def f(x):
   '''This function doubles its argument.

   Args:
      x: a number to double
   Returns:
      x*2
   '''
   y = x * 2
   return y
```

func36.py

The first few lines of the file document the module. These include a description of what the module does, and it's good programming practice to include the author's name and some indication of what version of the software it is (perhaps including the date). The first line of a function can also be a docstring. Here it is a good idea to describe the function and explain what arguments it takes and what values are returned.

2 They are used in a similar way to document classes and methods; see Chapter 9.

Docstrings are instances of public documentation. They can be displayed via the `help()` function whenever the module is loaded. For example, if we type the following at the prompt in the interactive environment, the docstrings for the module *and any public functions in that module* are displayed:

```
>>> import func36
>>> help(func36)
```

If your module includes private functions with a prefixed underscore _ that have docstrings, they will *not* be displayed with this call.

Your Python distribution should also include the `pydoc` command. You can use this at the command line to display the same information:

```
> pydoc ./func36.py
```

Notice that you have to specify the location of the module file in the call to `pydoc`. In this case `./` indicates that the file is in the current directory (on a Mac or Linux machine).

Note that this is not the same thing as commenting your code. Comments are something you use to remind you how you wrote your program, how the code is structured. Docstrings are there so that *other people* can use your code.

If somebody wants to use your modules and functions, they should be able to get what they need from the documentation you put in your docstrings. If you or somebody else wants to adapt your code, the docstrings will not be enough and your comments will also be necessary.

5.8 Analysis of Sentences

In this section we incrementally develop a larger program that does some superficial analysis of sentence structure. We will again use the *Alice* text we used in Chapter 4. The program will be modular in that we will break up our code into separate functions and modules.

We first write the framing code to read in the whole text. We print out the number of characters in the text to make sure we're actually reading the whole thing in. Note that we're reading the whole thing in here, rather than reading it in line by line, because we will ultimately be interested in sentence-sized units that do not correspond to line breaks.

```
f = open('alice.txt','r') #open file
text = f.read()           #read it all in
f.close()                 #close stream
```

```
#check that that worked!
print('characters:',len(text))
```

func37.py

Recall that files from Project Gutenberg begin with extra header information, which we want to remove. If we try various options, we'll see that that header comprises 10,840 characters. The following version of the code prints out those first 10,840 characters. We then remove that and print the first 100 characters of the remainder.

```
f = open('alice.txt','r') #open file
text = f.read()           #read whole text
f.close()                 #close file stream
print(text[:10840])       #print the header
text = text[10841:]       #remove the header
#something to separate two print statements
print('NEW START OF FILE:\n')
#print first 100 letters of what remains
print(text[:100])
```

func38.py

Let's begin modularizing the program. We'll factor out the file IO part as a separate function like this:

```
#function to read in file
#and prune header info
def readfile(filename):
   f = open(filename,'r')
   text = f.read()
   f.close()
   text = text[10841:]
   return text

#invoke the function
t = readfile('alice.txt')
#print the number of letters
#to make sure this worked
print('characters:',len(t))
```

func39.py

Let's now start a new function to split the text into sentences. We will see better ways to do this in Chapter 6. Our goal here is to develop a large modular program, so we set aside details of how best to split into sentences. First, we factor that part out with just a shell that does nothing:

```
#function to read in file
#and prune header info
def readfile(filename):
   f = open(filename,'r')
   text = f.read()
   f.close()
   text = text[10841:]
   return text

#function to split into sentences
def getsentences(t):
   return t

#read in file and strip header
txt = readfile('alice.txt')
#split into sentences (ultimately)
s = getsentences(txt)
#print first 10 sentences
for i in range(10):
   print(i,s[i])
```

func40.py

The function `getsentences()` does nothing at this point, but our first step in defining the function is to create a function that does nothing so that we can be sure that we're calling the function correctly. We'll now start fleshing it out. The next step is to just split the string on period. To do this, we use the `split()` method for strings.

```
#function to read in file
#and prune header info
def readfile(filename):
   f = open(filename,'r')
   text = f.read()
   f.close()
   text = text[10841:]
```

```
    return text

#function to split into sentences
def getsentences(t):
    ss = t.split('.')
    return ss

#read in file and strip header
txt = readfile('alice.txt')
#split into sentences
s = getsentences(txt)
#print first 10 sentences
for i in range(10):
    print(i,s[i])
```

func41.py

This is certainly moving in the right direction, but it falls short in two respects. First, it fails to separate sentences that terminate with something other than a period, e.g., question mark and exclamation point. Second, the `split()` method consumes the period, rather than terminating the string with it.

To overcome these shortcomings, we write our own string-splitting function.

```
#function to read file, prune header info
def readfile(filename):
    f = open(filename,'r')
    text = f.read()
    f.close()
    text = text[10841:]
    return text

#function to split into sentences
def getsentences(t):
    splitters = '.?!' #characters to split on
    ss = []           #put sentences here
    i = 0
    #go character by character
    while i < len(t):
        s = ''        #reset current sentence
        #read to end of text or
        #end of a sentence
```

```
        while i < len(t) and \
           t[i] not in splitters:
              s += t[i]
              i += 1
        #if text isn't over, current character
        #is splitter and should be appended
        if i < len(t):
           s += t[i]
        i += 1               #go on to next character
        ss.append(s)         #add sentence to list
     return ss

  #read in file and strip header
  txt = readfile('alice.txt')
  #split into sentences
  s = getsentences(txt)
  #print first 10 sentences
  for i in range(10):
     print(i,s[i])
```

func42.py

We've fleshed out the `getsentences()` function quite a bit and the logic is a bit complex, so let's go through it. The basic idea is to step through the text character by character until the end. This is the `while i < len(t)` loop. We also set up two variables. One, `ss`, is the list of sentences we find. As we find sentences, we add them to this list. The other variable is the current sentence `s`. Inside this loop, we examine the current character to test if it is a sentence-ending character. If it is not, we add it to the current sentence and go on to the next character. If, on the other hand, it is a sentence-ending character, we add it to the sentence, add the sentence to the list of sentences, reset the current sentence to `''`, and go on to the next character.

The logic is tricky, so we've commented the function extensively above. Note that we've had to spread the second `while` loop line over two lines. Python requires that you end the first half of the line with \.

The two functions we've written, `getsentences()` and `readfile()`, differ in their generality. The `readfile()` function is specific to the *Alice* text since it trims off a specific number of characters to eliminate the Gutenberg header. The `getsentences()` function, however, is more general. We can well imagine that we might use it again in some other program where we need to break a text into sentences.

On this reasoning, it's appropriate to put `getsentences()` into its own module that we can invoke whenever we need to split a text into sentences. This module is quite simple and merely includes the `getsentences()` function on its own:

```
#function to split into sentences
def getsentences(t):
   splitters = '.?!' #characters to split on
   ss = []           #put sentences here
   i = 0
   #go character by character
   while i < len(t):
      s = ''          #reset current sentence
      #read until the end of the text or
      #end of a sentence
      while i < len(t) and \
         t[i] not in splitters:
            s += t[i]
            i += 1
      #if text isn't over, current character
      #is splitter and should be appended
      if i < len(t):
         s += t[i]
      i += 1          #go on to next character
      ss.append(s)    #add sentence to list
   return ss
```

func43.py

We can eliminate the function definition from our main program and import it from `func43.py`:

```
import func43

#function to read in file
#and prune header info
def readfile(filename):
   f = open(filename,'r')
   text = f.read()
```

```
    f.close()
    text = text[10841:]
    return text

#read in file and strip header
txt = readfile('alice.txt')
#split into sentences
s = func43.getsentences(txt)
#print first 10 sentences
for i in range(10):
    print(i,s[i])
```

func44.py

Note that if we truly want to make this new module generally useful, we would name it something appropriate, something mnemonic, e.g., `splitter.py` or the like. We've chosen to use the name `func43.py` to keep all the program files organized.

Note now that the sentences we are printing out have odd spaces, tabs, and returns interspersed. In particular, there are returns and tabs in each sentence. In addition, there are instances of multiple spaces. Let's clean these up. Specifically, we'll convert all returns and tabs to spaces. We'll then take all instances of multiple spaces in a row and convert them to a single space. Finally, we'll eliminate any spaces at the beginning or end of a line. The following program includes a new function, `makespaces()`, that converts returns and tabs to spaces and then converts any sequence of spaces to a single space.

```
import func43

#function to read in file, prune header info
def readfile(filename):
    f = open(filename,'r')
    text = f.read()
    f.close()
    text = text[10841:]
    return text

#remove non-space breaks and trim spaces
def makespaces(t):
    breaks = '\n\t'      #characters to convert
    r1 = ''              #output of 1st convert
    i = 0
```

```
    while i < len(t):  #go through 1 by 1
        #current char should be converted?
        if t[i] in breaks:
            r1 += ' '
        else:
            r1 += t[i]
        i += 1
    #eliminate space after another space
    r2 = r1[0]
    i = 1                #start at 2nd char
    #go through whole thing
    while i < len(r1):
        #check for two spaces in a row
        if r1[i] == ' ' and \
            r2[len(r2)-1] == ' ':
                #skip if so
                i += 1
                continue
        #otherwise, append current char
        else:
            r2 += r1[i]
        i += 1
    return r2

#read in file and strip header
txt = readfile('alice.txt')
#remove non-space breaks and trim spaces
cleanedtext = makespaces(txt)
#split into sentences
s = func43.getsentences(cleanedtext)
#print first 10 sentences
for i in range(10):
    print('\n',i,': ',s[i],sep='')
```

func45.py

The logic of this new function is as follows. First, we set up an empty output string `r1`. The function then goes through the text character by character. If the current character is a return or tab, then it appends a space to `r1`. Otherwise, the current character is appended. We then set up another output string `r2` with just the first character of `r1`. The function then goes through `r1` starting at the second character. If the current character of `r1` is a space and the last character of `r2` is a space, it is skipped; otherwise, the current character of `r1` is appended to `r2`. We then return `r2`.

We now need a function to trim extra spaces on the edges of each sentence. The following program adds this:

```
import func43

#function to read in file, prune header info
def readfile(filename):
    f = open(filename,'r')
    text = f.read()
    f.close()
    text = text[10841:]
    return text

#remove non-space breaks and trim spaces
def makespaces(t):
    breaks = '\n\t'       #characters to convert
    r1 = ''               #output of 1st convert
    i = 0
    while i < len(t):     #go through 1 by 1
        #current char should be converted?
        if t[i] in breaks:
            r1 += ' '
        else:
            r1 += t[i]
        i += 1
    #eliminate space after another space
    r2 = r1[0]
    i = 1                 #start at 2nd char
    #go through the whole thing
    while i < len(r1):
        #check for two spaces in a row
        if r1[i] == ' ' and \
            r2[len(r2)-1] == ' ':
                i += 1    #skip if so
                continue
        #otherwise, append current char
        else:
            r2 += r1[i]
```

```
      i += 1
   return r2

#remove spaces at edges of strings
def trimspaces(t):
   r1 = []              #result list
   for s in t:          #go through 1 by 1
      #if first char is a space
      if s[0] == ' ':
         s = s[1:]
      slast = len(s) - 1
      #if last char is a space
      if len(s) > 0 and s[slast] == ' ':
         s = s[:slast]
      r1.append(s)
   r2 = []              #prune empty sentences
   for s in r1:         #go through one by one
      #check if sentence is empty
      if len(s) > 0:
         r2.append(s)
   return r2

#read in file and strip header
txt = readfile('alice.txt')
#remove non-space breaks and extra spaces
cleanedtext = makespaces(txt)
#split into sentences
ss = func43.getsentences(cleanedtext)
#trim edges of sentences
ts = trimspaces(ss)
#print first 10 sentences
for i in range(10):
   print('\n',i,': %',ts[i],'%',sep='')
```

func46.py

The new `trimspaces()` function has two loops. The first goes through and eliminates an initial or final space on all strings. This can now result in empty strings, so the second loop eliminates these. In both cases, we do this by creating new lists of strings and only transferring sentences to these new lists if they satisfy the properties we specify. Notice how we've changed the call to `print()` at the end so we can see the effects of our string-trimming function.

The two functions we've added, `makespaces()` and `trimspaces()`, are both general, so we will move them both to our separate module. Here is the revised module:

```
#function to split into sentences
def getsentences(t):
   splitters = '.?!' #characters to split on
   ss = []           #where we put sentences
   i = 0
   #go character by character
   while i < len(t):
      s = ''         #reset current sentence
      #read to end of text or end of sentence
      while i < len(t) and \
         t[i] not in splitters:
            s += t[i]
            i += 1
      #if text isn't over, current character
      #is splitter and should be appended
      if i < len(t):
         s += t[i]
      i += 1         #go on to next character
      ss.append(s)   #add current sentence
   return ss

#remove non-space breaks and trim spaces
def makespaces(t):
   breaks = '\n\t'   #characters to convert
   r1 = ''           #output of 1st convert
   i = 0
   #go through 1 by 1
   while i < len(t):
      #current char should be converted?
      if t[i] in breaks:
         r1 += ' '
      else:
         r1 += t[i]
      i += 1
   #eliminate space after another space
   r2 = r1[0]
```

```
    i = 1              #start at 2nd char
    #go through the whole thing
    while i < len(r1):
        #check for two spaces in a row
        if r1[i] == ' ' and \
            r2[len(r2)-1] == ' ':
                i += 1    #skip if so
                continue
        #otherwise, append current char
        else:
            r2 += r1[i]
        i += 1
    return r2

#remove spaces at edges of strings
def trimspaces(t):
    #result list
    r1 = []
    #go through one by one
    for s in t:
        #if first char is a space
        if s[0] == ' ':
            s = s[1:]
        slast = len(s) - 1
        #if last char is a space
        if len(s) > 0 and s[slast] == ' ':
            s = s[:slast]
        r1.append(s)
    #prune empty sentences
    r2 = []
    #go through one by one
    for s in r1:
        #check if sentence is empty
        if len(s) > 0:
            r2.append(s)
    return r2
```

func47.py

Here is the revised program that now calls three functions from `func47.py`:

```
import func47

#function to read in file, prune header info
def readfile(filename):
   f = open(filename,'r')
   text = f.read()
   f.close()
   text = text[10841:]
   return text

#read in file and strip header
txt = readfile('alice.txt')
#remove non-space breaks and extra spaces
cleanedtext = func47.makespaces(txt)
#split into sentences
ss = func47.getsentences(cleanedtext)
#trim edges of sentences
ts = func47.trimspaces(ss)
#print first 10 sentences
for i in range(10):
   print('\n',i,': %',ts[i],'%',sep='')
```

func48.py

With all this as background, we can now add the functionality we're really interested in. What is the distribution of sentences in terms of average number of words? The code here is actually quite straightforward and similar to the final bit of code in `io27.py` in Section 4.4.

```
import func47

#function to read in file, prune header info
def readfile(filename):
   f = open(filename,'r')
   text = f.read()
   f.close()
   text = text[10841:]
   return text

#read in file and strip header
txt = readfile('alice.txt')
#remove non-space breaks and extra spaces
```

```
cleanedtext = func47.makespaces(txt)
#split into sentences
ss = func47.getsentences(cleanedtext)
#trim edges of sentences
ts = func47.trimspaces(ss)
#dictionary to keep track of counts
counts = {}
#go through all sentences
for s in ts:
   slength = len(s.split())  #count words
   #add 1 to relevant count in dictionary
   if slength in counts:
      counts[slength] += 1
   else:
      counts[slength] = 1
for c in sorted(counts):       #print counts
   print(c,counts[c])
```

func49.py

Here we define a dictionary, `counts`, to keep track of the number of occurrences of sentences of different lengths. We iterate through all the sentences adding to `counts` as appropriate. Finally, we print out all values of `counts`. We sort the keys of the dictionary so they are in order.

5.9 Exercises

5.1 What does this function do?

```
def f(x):
   return x
```

5.2 What happens if we apply the function above like this: `f(f)(f(3))`? Explain.

5.3 What does the following code do? Explain.

```
(lambda f: f(3))(lambda x: x+2)
```

5.4 Write a function that *composes* two other functions. That is, it takes two functions as arguments and returns a function that applies the first function and then the second function to its argument. The two functions should each take a single argument. (To be clear, your function should return the composed function, not the result of the composition.)

5.5 Write a function that has the same effect as `*` in an expression like:

```
'this' * 3
```

You are not allowed to use `*` in your function definition!

5.6 Write a function that is helpful to you. It should take at least *three* arguments.

5.7 Add appropriate docstrings to `func47.py` and `func49.py`.

5.8 Write a function that takes a filename and a number *n* as arguments. The function will open the file and return the first *n* words of the file. Write the function so that it recovers from errors: what happens if the file doesn't exist, what if the number isn't a number, what if the file isn't long enough to return *n* words, etc.

5.9 Write a function that takes two string arguments. The first argument is a string to search for and count instances of the second argument. Thus, if the function were invoked with `'Mississippi'` and `'is'`, it would return `2`. The function must be recursive. Make sure your function can handle cases where the second argument is of any length.

5.10 **Web**: Snoop around on the web and find a publically available module that does something useful for you. Explain what it is and write a short program demonstrating its utility.

6 Regular Expressions

In this chapter we discuss one of the most important aspects of programming for linguists: pattern matching. You will often have to assess whether a string matches a pattern, contains a certain sequence of characters or a specific number of characters, etc.

We've already seen that we can do this with what we've already learned. For example, if we wanted to know if a string s contained the sequence ab, we could simply write 'ab' in s, which would return True or False. If, on the other hand, we wanted to know if a string contained 'a' and then 'b' with material potentially intervening, we would have to do more. We might write a function like this:

```
import sys

def mymatch(s):          #a and then b
   i = 0
   #flag to keep track of
   #whether we see an 'a'
   aFlag = False
   while i < len(s):
      if s[i] == 'a':
         aFlag = True
         break
      i += 1
   #look for 'b' where we left off
   while i < len(s):
      if s[i] == 'b':
         #if we find 'b', return True of
         #False depending on whether we
         #previously saw 'a'
         return aFlag
      i += 1
```

```
    #if all that fails, return False
    return False

print(mymatch(sys.argv[1]))
```

re1.py

The function `mymatch()` can be invoked with a command-line argument, so we can test out how it applies to different strings. The logic of the function is that we search the string for the first instance of `'a'` by iterating across the string with the counter `i`. If we find it, we set the value of `aFlag` to `True` and exit the first `while` loop. Then, without resetting `i`, we initiate another `while` loop to look for `'b'`. Not resetting the value of `i` means that the second loop picks up where the first ended. Hence, the `'b'` must follow the (first) `'a'`.

This is fine as far as it goes, but it doesn't generalize. There are an infinite number of patterns we might want to search for, and writing a potentially complex matching function for each one is at best tedious and at worst can lead to programming errors.

Many programming languages, Python included, implement a general pattern matching mechanism that is flexible and efficient: *regular expressions*. This is not a book about computational linguistics generally, so we don't have time to go into it fully here, but suffice it to say that the theory behind regular expressions is both fascinating and powerful. We reluctantly set it aside here and focus on practicalities: how pattern matching with regular expressions works in Python.

The organization of this chapter is as follows. We first lay out basic pattern-patching functions in Python. We then turn to how patterns can be specified using regular expressions. We then discuss pattern-matching devices that go beyond regular expressions. Finally, we give an extended example showing how pattern matching with regular expressions can be used for linguistic purposes.

6.1 Matching

Matching a string against some pattern is done with the `re` module, and most typically with the `search()` function in that module. At its simplest, we can invoke it like this:

```
import re,sys

if re.search('ab',sys.argv[1]):
    print('a match')
```

```
else:
   print('no match')
```

re2.py

Here we import from the sys module to make use of command-line arguments and from re to use the pattern-matching function search(). This function takes two arguments: a pattern and a string. In this case, the pattern 'ab' matches any string that contains that letter sequence, e.g., 'abc', 'xab', 'ab', 'xabc'.[1]

We'll consider the syntax of patterns in depth in the next section, but let's consider one slightly more complex pattern here: 'a.*b'. This matches a string that contains 'a' followed anywhere by a 'b'. You can see this by trying different command-line arguments with the following program:

```
import re,sys

if re.search('a.*b',sys.argv[1]):
   print('a match')
else:
   print('no match')
```

re3.py

The search() function actually does not return True or False, but rather a match object if the string matches or None if it does not. The reason why the code in the two programs above works is that in an if statement, a match object will evaluate to True and a None object will evaluate to False. The following code shows this clearly.

```
import re

#do two matches
res1 = re.search('a.*b','hat')
res2 = re.search('a.*b','nab')

#evaluate results of both matches
for s,r in [('hat',res1),('nab',res2)]:
```

[1] The re.search() function can take an additional argument flags, which can take a number of values. The only ones you are likely to use are flags=re.I, which allows case-insensitive matching, and flags=re.S, which allows . to match a return in multiline situations. If you use them both, they must have a pipe in between: flags=re.S|re.I. We exemplify this below.

```
    if r:                  #simple if test
       print(s,"matches 'a.*b'")
       print("r is a match object")
    else:
       print(s,"does not match 'a.*b'")
       print("r is None")
    if r == False:         #does match simply fail?
       print('r == False')
    else:
       print('r != False')
    if r == None:          #is match a None object?
       print('r == None')
    else:
       print('r != None')
```

re4.py

Here we explicitly save the output of the `search()` function and compare it against `False` and against `None`.

In fact, a match object does more for us than evaluate to `True`. It is an object with several methods we can make use of: `group()`, `start()`, `end()`, and `span()`.

The `group()` method simply returns the matched part of the string. In the case of a pattern like `'a'`, this will simply return `'a'` in the case of a match. In the case of a more interesting pattern like `'a.*b'`, the `group()` method returns the entire string from `'a'` to `'b'`. The following code exemplifies:

```
import re,sys

#do a match
res = re.search('a.*b',sys.argv[1])
if res:    #if match succeeds, print matching
   print("match: '",res.group(),"'",sep='')
else:
   print('no match')
```

re5.py

If you try this with different command-line argument choices, you'll see that the `group()` method returns the entire string from `'a'` to `'b'`.

The `start()`, `end()`, and `span()` methods return the starting and ending indices of the matched portion. The `span()` method returns both. The following code exemplifies:

```
import re,sys

#do a match
res = re.search('a.*b',sys.argv[1])
if res: #if match succeeds, print everything
   print("match: '",res.group(),"'",sep='')
   print('starting index:',res.start())
   print('ending index:',res.end())
   print('both indices:',res.span())
else:
   print('no match')
```

re6.py

Notice that, as with expressions with :, the ending index is the index of the character *after* the final character of the pattern.

These methods allow us to see some other aspects of the code. First, the match begins at the earliest possible point in the string. In the case at hand, if there are multiple instances of `'a'`, the match begins with the first one. Second, the match is as *greedy* as possible. If there are multiple instances of `'b'`, the match uses the rightmost one.

Another useful general matching function is `findall()`. What this does is return all instances of the match in a string. For example, if you match `'a'` against `'abracadabra'`, this returns `['a','a','a','a','a']`. The following code exemplifies:

```
import re,sys

#find all matches
res = re.findall('a',sys.argv[1])
if res:                           #if at least 1
   print('match:',res)            #print list
else:
   print('no match:',res)   #print empty list
```

re7.py

Note that in the event of no match, `findall()` returns an empty list rather than `None`.

6.2 Patterns

In this section, we present the general structure and specific syntax of the kinds of patterns we can specify for the functions covered in the previous section and those to follow in the next chapter.

Patterns to be matched are defined in terms of *regular expressions* (REs). These have a simple syntax that we can specify recursively, like a phrase-structure grammar.[2]

(i) A single symbol is an RE. For example: `'a'`, `'3'`, `'k'`, etc. These will match a string that consists of just the symbol indicated.

(ii) A *concatenation* or sequence of REs is an RE. For example: `'ab'`, `'3g'`, `'kk'`, etc. Such an expression matches if the expressions it is composed of match in sequence. Thus `'ab'` matches if `'a'` matches and then `'b'` matches. Note that the definition is recursive, so if `'ab'` is legal and `'cd'` is legal, then so are `'abcd'` and `'cdab'`.

(iii) The *union* of disjunction of two REs is a regular expression. Union can be indicated with a tie-bar, e.g., in `'a|b'`, `'a|d'`. An expression like this is matched if either one or the other of the component expressions matches. Thus, for example, `'a|b'` matches if either `'a'` or `'b'` matches. This definition is recursive as well, so we can build up expressions with both union and concatenation repeatedly. This can result in ambiguity, and we can use parentheses to disambiguate. For example, in principle, something like `'a|bc'` is ambiguous and can be rewritten as `'(a|b)c'` or `'a|(bc)'`.

(iv) Finally, we have *Kleene star*. This allows an RE to be matched zero or more times. It is indicated with a following asterisk and is recursive as well. We have, for example: `a*`, `a(b*)`, `(ab)*`, `(a|b)*`, etc.

Everything else about REs can be reduced to these simple operations, so let's take some time to understand them in more depth.

An RE is a string defined in terms of the recursive operations above: concatenation, union, and Kleene star. An RE itself is a finite sequence of symbols, but it defines a potentially infinite set of strings. For example, the RE `ab*c` is a finite sequence of symbols, but defines an infinite set: $\{\text{ac}, \text{abc}, \text{abbc}, \text{abbbc}, \ldots\}$. Similarly, `(ab)|(c*)` defines the infinite set $\{\text{ab}, \epsilon, \text{c}, \text{cc}, \text{ccc}, \ldots\}$. We use ϵ to indicate the empty string.

The general idea in pattern matching is that if the string we are matching against contains any of the strings that the relevant RE defines, then there is a match. For example, if we try to match the RE `a|b` against the string `pancake`, we have a match because the RE defines the string set $\{\text{a}, \text{b}\}$ and `pancake` contains the substring `a`. Similarly, the RE `a*` will match any string

[2] Formally, one can show that these are more restricted than a phrase-structure grammar.

because the set of strings it defines includes ϵ and every string definitionally includes the empty string.

The following chart gives examples of simple REs (along the left) and whether they match various strings (along the top).

	`a`	`b`	`ab`	`acb`	`ba`
`a`	✓		✓	✓	✓
`ab`			✓		
`a|b`	✓	✓	✓	✓	✓
`a*`	✓	✓	✓	✓	✓
`a|(bc)`	✓		✓	✓	✓
`(a|b)c`				✓	
`a*|b`	✓	✓	✓	✓	✓
`a|(b*)`	✓	✓	✓	✓	✓
`(a|b)*`	✓	✓	✓	✓	✓
`a(b*)`	✓		✓	✓	✓
`(ab)*`	✓	✓	✓	✓	✓

Most programming languages use REs to do pattern matching because by using these it is possible to be extremely efficient when checking whether some string matches some pattern. If we were to enrich our pattern-matching system significantly beyond the three operations listed above, we would lose this efficiency.

There is a tradeoff, however. With this efficient system, there are patterns we cannot specify. Most programming languages, Python included, thus go slightly beyond REs in what they allow in pattern-matching syntax. The resulting system is still limited in what it allows, but for most purposes is more than enough for the kinds of pattern matching and string manipulation linguists need.

In the remainder of this section, we will go through some additional notational conventions available in REs. In most cases, these are for convenience and do not extend the power of REs.

To play with these, we'll make use of the following program. This program takes one command-line argument that specifies a pattern. That pattern is then matched against the words in the *Alice* text and printed to the screen.

```
import sys,re
#read in 'Alice' and break into words
f = open('alice.txt','r')
words = f.read().split()
f.close()
for w in words:  #match against each word
```

```
    m = re.search(sys.argv[1],w)
    if m:
        print(w)
```

re8.py

For example, we might invoke it like this:

```
> python re8.py ab*c
...
```

We will see below that some of the special characters we use in REs also have an interpretation in some operating systems. We must therefore enter the pattern in quotes or use other special conventions to make sure the pattern is passed through the operating system to Python. We note this where relevant below.

We've already seen the notations for concatenation, union, and Kleene star. We've also seen that parentheses can be used to disambiguate.

A simple addition to these notations is `'.'`, which matches any single character. Thus an expression like `'a.b'` will match any string where `'b'` follows `'a'` with a single character intervening. Keep in mind that that single characters can be alphabetic, numeric, space, tab, etc. For example, `'...'` matches any string with at least three letters.

We can add `'^'` and `'$'` as well. The former matches the beginning of the string, and the latter matches the end of the string. Thus `'ab'` matches any string that contains that substring, but `'^ab'` matches only strings that begin with that substring. Similarly, `'ab$'` will match any string that ends with that substring. A pattern like `'^ab$'` will only match that exact string. The RE `'^...$'` will match any string that has exactly three characters.

We've already seen that tie-bar can be used to indicate disjunction or union. For example, something like `'(a|b|cd|ef)g'` will match a string that contains any of these as substrings: `'ag'`, `'bg'`, `'cdg'`, or `'efg'`. If the elements in the union are single characters, then Python offers an alternative notation using square brackets instead of the tie-bar. For example, something like `'a(b|c|d)e'` can also be written `'a[bcd]e'`. This is *not* possible if any of the elements in the union is more than a single character, like the first example in this paragraph.

Interestingly, the square bracket syntax – but not the tie-bar – can be used to specify the *inverse* set of characters. The syntax is to put the caret symbol first inside the square brackets to denote the opposite set of characters. Thus `'[^abc]'` will match a single character other than `'a'`, `'b'`, or 'c'. As with simple square brackets, this works only with single individual characters.

Either square bracket expression can be used with a naturally ordered character sequence denoted with hyphen. Thus [a-e] is the same as [abcde], and [0-5] is the same as [012345]. Multiple sequences can be used in the same square brackets. For example, [a-zA-Z] denotes a single upper- or lowercase letter. Finally, sequences can be used in inverse character classes. Hence [^a-z] denotes a single character that is not a lowercase letter.

We've already discussed Kleene star, which lets an expression be matched zero or more times. Thus (b|c)* matches zero or more instances of b or c. There are two related notations: (b|c)+ indicates one or more instances of b or c, and (b|c)? indicates zero or one instance of b or c.

There is a whole set of character classes that are predefined. The most useful of these include \w "word" characters, \W "nonword" characters, \s whitespace, \S nonwhitespace, \d digits (equivalent to [0-9]), and \D nondigits (equivalent to [^0-9]). (Note that all of these must occur within quotes when used on the command line.)

6.3 Backreferences

Backreferences technically move pattern matching beyond regular expressions. While they provide a lot of power and convenience, they should be used reluctantly, as they can significantly affect how efficiently your programs run.

The simplest characterization of backreferences is that every time you use parentheses in a pattern, you can refer back to them later. These can be parentheses you've used to disambiguate a pattern, e.g., a|(bc) versus (a|b)c, but they can also be parentheses you've simply added with no such effect, e.g., .*a versus (.*)a.

One way you can refer back to them is with a match object. If you've matched a pattern with parentheses, then you can extract that portion from the match object with the group(), start(), end(), or span() methods. The following program exemplifies:

```
import sys,re

#do a match
m = re.search('(.*)b(.*)',sys.argv[1])
if m:                        #if it succeeds, print...
    #the whole match
    print('all: "',m.group(),'"',sep='')
    #the first part
    print('group 1: "',m.group(1),'"',sep='')
```

```
    #the second part
    print('group 2: "',m.group(2),'"',sep='')
```

re9.py

In this program, we invoke the `group()` method with an integer argument. The integer refers back to parentheses in the pattern. For example, in an expression with three sets of parentheses, we can refer back to any of `group(1)`, `group(2)`, or `group(3)`. In fact, `group()` is the same as `group(0)`. The first set of parentheses in `re9.py` are of this gratuitous sort.

Note, incidentally, that the parentheses don't have to be justified to disambiguate. Gratuitous parentheses that don't disambiguate the pattern can also be used and referred back to with `group()`.

The other match object methods can also be used with backreferences. The following code shows how this works.

```
import sys,re

m = re.search('(.*)b(.*)',sys.argv[1])
if m:
    print('all: "',m.group(),'"',sep='')
    print('group 1: "',m.group(1),'"',sep='')
    print('group 1 start:',m.start(1))
    print('group 1 end:',m.end(1))
    print('group 2: "',m.group(2),'"',sep='')
    print('group 2 start:',m.start(2))
    print('group 2 end:',m.end(2))
```

re10.py

Here, we print out the start and end indices for groups 1 and 2.

Interestingly, there is a notation that allows us to use backreferences in the same pattern: `\1`, `\2`, `\3`, etc. For example, `'(..)\1'` will match any two-letter sequence that is repeated, e.g., `abab`, `bcbc`, `bbbb`. This is an extremely powerful notation and can move pattern matching beyond simple regular expressions. Use this one with caution.

6.4 Initial Consonant Clusters

Let's now exemplify how to use pattern matching to investigate the distribution of word-initial consonant clusters in *Alice*. The basic question is how frequent different kinds of initial consonant clusters are.

A major challenge here is that English orthography only indirectly reflects the phonology. We will therefore set aside some issues about which precise cluster different letter sequences represent. There are several sorts of problems to deal with. First, we must deal with silent letters. These include cases like *know* or *gnostic* where *k* and *g* are silent. Another issue to wrestle with is that sometimes a single consonant is written with several letters, e.g., [θ] as in *thimble* or [ʃ] as in *show*. Finally, there are cases where the same letter (sequence) has multiple pronunciations. For example, *c* is pronounced [s] in *city*, but [k] in *coat*. Similarly, *ch* is pronounced [tʃ] in *church*, but [k] in *chord*. Note that sometimes the difference is predictable, as in the case of *c*, but sometimes it is not, as in the case of *ch*. We will therefore set aside the mapping of orthography to precise phonological forms, except insofar as we need to decide whether something is or is not a cluster.

We will, as always, build our code up incrementally. Let's first write the framing code to read in the *Alice* text.

```
f = open('alice.txt','r') #open the file
text = f.read()            #read it all in
f.close()                  #close file stream
#print first 100 letters to make sure
print(text[:100])
```

re11.py

We print out the first 100 characters to make sure everything is working.

Anticipating our task, we strip the Gutenberg header, convert the whole text to lowercase, and then split it into words.

```
f = open('alice.txt','r') #open file
text = f.read()            #read it all in
f.close()                  #close file stream
text = text[10841:]        #remove header
#convert to lowercase and split into words
words = text.lower().split()
#print first 50 words
for w in words[:50]:
   print(w)
```

re12.py

Again, we print out the first 50 words to make sure things are working right.

Our first challenge is nonalphabetic characters. To see what the problem is more specifically, we can make use of the `re8.py` program we wrote above to see what kinds of nonalphabetic things we're getting. We do this with the following command:

```
> python re8.py '\W'
```

We see immediately that many of these are cases where some punctuation like comma or period is on the right end of the word. Let's exclude these cases and see what's left. We therefore try this:

```
> python re8.py '\W\w'
```

Now what we see are mostly apostrophes, hyphens, and single and double quotation marks. We must clearly exclude initial quotation marks; we don't want these to confuse what we take to be an initial consonant cluster.

Hyphens are another matter. It looks like a double hyphen is essentially a word break; the most reasonable thing would be to convert those to a single space. Single hyphens as in *jury-box* are more complex. Do we want to consider *box* here in our calculation of initial clusters? This is effectively a theoretical decision. We'll therefore simply choose to include cases like *box* in *jury-box* as items that have initial clusters. On this assumption, a single hyphen needs to be converted to a space as well.

The simplest thing is to simply convert all of these to spaces before we break the text into words. The following code handles this:

```
import re

f = open('alice.txt','r')  #read in Alice
text = f.read()
f.close()
text = text[10841:]         #strip header
#convert to lower case
lowertext = text.lower()
#punctuation to convert
punc = '[\.\?\-!\?\*,"\(\):\`\[\];_/~]'
#convert punctuation to space
newtext = ''
for c in lowertext:
   if re.search(punc,c):
      newtext += ' '
   else:
```

```
        newtext += c
words = newtext.split()     #split into words
#print first 50 words
for w in words[:50]:
    print(w)
```

re13.py

There are now apostrophes and single quotes to deal with: these we delete. The following revision does this:

```
import re

f = open('alice.txt','r')  #read in Alice
text = f.read()
f.close()
text = text[10841:]         #strip header
#convert to lower case
lowertext = text.lower()
#punctuation to convert
punc = '[\.\?\-!\?\*,"\(\):\`\[\];_/~]'
#convert punctuation to space
newtext = ''
for c in lowertext:
    if re.search(punc,c):
        newtext += ' '
    else:
        newtext += c
words = newtext.split()     #split into words
#delete single quotes
newwords = []
for w in words:
    word = ''
    for c in w:
        if c != "'":
            word += c
    newwords.append(word)
#print first 50 words
for w in newwords[:50]:
    print(w)
```

re14.py

This last version is a little complex, as we're constructing a new list `newwords` by going through the old list word by word and constructing a new word `word` by going through each old word letter by letter.

Finally, we can see that there are several words that either are numbers or contain numbers. We'll just eliminate these altogether.

```
import re

f = open('alice.txt','r')   #read in Alice
text = f.read()
f.close()
text = text[10841:]         #strip header
#convert to lower case
lowertext = text.lower()
#punctuation to convert
punc = '[\.\?\-!\?\*,"\(\):\`\[\];_/~]'
#convert punctuation to space
newtext = ''
for c in lowertext:
   if re.search(punc,c):
      newtext += ' '
   else:
      newtext += c
#split into words
words = newtext.split()
#eliminate single quotes
newwords = []
for w in words:
   word = ''
   for c in w:
      if c != "'":
         word += c
   newwords.append(word)
#eliminate words with numbers
finalwords = []
for w in newwords:
   if re.search('[0-9]',w):
      continue
   else:
      finalwords.append(w)
#print first 50 words
```

```
for w in finalwords[:50]:
   print(w)
```

re15.py

Our next step is to identify consonant clusters. We've already seen that the identification of certain consonants is lexical, that the identification of what a letter stands for sometimes is based on what word it occurs in. For example, *g* is [g] in *get*, but [dʒ] in *gem*.

We will therefore aggregate by words before doing counts for individual clusters. The following code does this:

```
import re

f = open('alice.txt','r')  #read in Alice
text = f.read()
f.close()
text = text[10841:]        #strip header
#convert to lower case
lowertext = text.lower()
#punctuation to convert
punc = '[\.\?\-!\?\*,"\(\):\`\[\];_/~]'
#convert punctuation to space
newtext = ''
for c in lowertext:
   if re.search(punc,c):
      newtext += ' '
   else:
      newtext += c
#split into words
words = newtext.split()
#eliminate single quotes
newwords = []
for w in words:
   word = ''
   for c in w:
      if c != "'":
         word += c
   newwords.append(word)
#eliminate words with numbers
finalwords = []
for w in newwords:
```

```
      if re.search('[0-9]',w):
         continue
      else:
         finalwords.append(w)
#do counts for words
wordlist = {}
for w in finalwords:
   if len(w) > 0:
      if w in wordlist:
         wordlist[w] += 1
      else:
         wordlist[w] = 1
#sort the words
keys = sorted(wordlist.keys())
#print out the first 100 words
for i in range(100):
   print(keys[i],wordlist[keys[i]])
#print out the number of distinct words
print('Keys:',len(keys))
```

re16.py

This code makes use of a new function, `sorted()`, which sorts a list of strings (or numbers). Note that we've chosen to aggregate the words into a list before converting to a dictionary here. This allows us to alter that list in various ways *before* doing counts.

Let's now begin to modularize our code. First, we convert `re16.py` into a module `re17.py` that we can call.

```
import re

def preprocess():
   f = open('alice.txt','r')  #read in Alice
   text = f.read()
   f.close()
   text = text[10841:]        #strip header
   #convert to lower case
   lowertext = text.lower()
   #punctuation to convert
   punc = '[\.\?\-!\?\*,"\(\):\`\[\];_/~]'
   #convert punctuation to space
   newtext = ''
   for c in lowertext:
```

```
        if re.search(punc,c):
           newtext += ' '
        else:
           newtext += c
    #split into words
    words = newtext.split()
    #eliminate single quotes
    newwords = []
    for w in words:
        word = ''
        for c in w:
           if c != "'":
              word += c
        newwords.append(word)
    #eliminate words with numbers
    finalwords = []
    for w in newwords:
        if re.search('[0-9]',w):
           continue
        else:
           finalwords.append(w)
    #do counts for words
    wordlist = {}
    for w in finalwords:
        if len(w) > 0:
           if w in wordlist:
              wordlist[w] += 1
           else:
              wordlist[w] = 1
    return wordlist
```

re17.py

That module contains a single function `preprocess()` that we can call from other programs like this:

```
import re17

wordlist = re17.preprocess()
#sort the words
keys = sorted(wordlist.keys())
#print out the first 100 words
for i in range(100):
```

```
   print(keys[i],wordlist[keys[i]])
#print out the number of distinct words
print('Keys:',len(keys))
```

re18.py

The first order of business is to find all possible word-initial consonant clusters. Here's a first pass just at getting this list:

```
import re,re17

#get the word counts
wordlist = re17.preprocess()
#just get the words
words = wordlist.keys()
clusters = []          #strip off onsets
for w in words:
   m = re.search('^[^aeiou]*',w)
   if m:
      onset = w[0:m.end()]
      clusters.append(onset)
#eliminate duplicate onsets
clusters = sorted(set(clusters))
for c in clusters:    #print all onsets
   print("'",c,"'",sep='')
print(len(clusters)) #print number of onsets
```

re19.py

This program makes use of the `set()` function, which converts a list to a set, the effect of which is to remove duplicates from the list. We then use `sorted()` to sort it alphabetically. This results in the following 76 hypothetical onset clusters.

null	b	bl	br	by	c	ch	chr	chrys
cl	cr	cry	d	dr	dry	f	fl	fly
fr	fry	g	gl	gr	gryph	h	hjckrrh	hm
j	k	kn	l	ly	m	my	mys	myst
n	p	pl	pr	q	r	s	sc	sch
scr	sh	shr	shy	shyly	sk	sky	sl	sm
sn	sp	spl	spr	sq	st	str	sw	t
th	thr	tr	try	tw	v	w	wh	why
wr	x	y	z					

There's a fair amount of noise here that we need to clean up. We can track down some of it by using our `re8.py` program above (remembering that we converted to lowercase and removed a fair amount of punctuation).

A big source of noise is the treatment of *y* and other vowels. Specifically, if we require words to contain a vowel and allow *y* to count as a vowel when it is not word-initial, we can eliminate many of these. The following revision does this:

```
import re,re17

#get the word counts
wordlist = re17.preprocess()
words = wordlist.keys()   #just get the words
clusters = []             #strip off onsets
for w in words:
   m = re.search('^([^aeiouy]*)[aeiouy]',w)
   if m:
      if m.end(1) == 0 and w[0] == 'y':
         onset = 'y'
      else:
         onset = w[0:m.end(1)]
      clusters.append(onset)
#eliminate duplicate onsets
clusters = sorted(set(clusters))
for c in clusters:        #print all onsets
   print("'",c,"'",sep='')
#print number of onsets
print(len(clusters))
```

re20.py

This now produces a reasonable list of 58 hypothetical word onsets.

Our last step is to do counts for all clusters. The following code does this:

```
import re,re17

#get the word counts
wordlist = re17.preprocess()
#just get the words
words = wordlist.keys()
#strip off onsets and do counts
clusters = {}
```

```
for w in words:
   m = re.search('^([^aeiouy]*)[aeiouy]',w)
   if m:
      if m.end(1) == 0 and w[0] == 'y':
         ons = 'y'
      else:
         ons = w[0:m.end(1)]
      if ons in clusters:
         clusters[ons] += 1
      else:
         clusters[ons] = 1
#print onset counts
keys = sorted(clusters.keys())
for c in keys:
   print("'",c,"': ",clusters[c],sep='')
```

re21.py

6.5 Exercises

6.1 Write a function that will match a string that contains `a.*b.*c.*d` *without* using regular expressions.

6.2 What will the pattern `'..^'` match? Explain why.

6.3 What's the difference between `(a*)|(b*)` and `(a|b)*`? Explain the difference and give examples of strings that both will match, only the first will match, and only the second will match.

6.4 Describe in words what each of the following patterns will match:

(a) `'(Tom)|(Dick)|(Harry)'`
(b) `'(.*)\1'`
(c) `'(a[^a])+'`
(d) `'^[^0-9]+$'`
(e) `'(a|b)*c(d*|e*)'`

6.5 Write a program that collects the *final* consonant clusters of English.

6.6 Write a regular expression that finds reduplicated words.

6.7 Write a program that uses regular expressions to find palindromes.

6.8 Some of the work of the string method `format()` can be done with regular expressions instead. Write a function that will take arguments where the first argument is a string in the appropriate format for the `format()` method and subsequent arguments are the additional ones that `format()` would take. Your function should, insofar as your patience takes you, do the work of `format()`.

6.9 For LaTeX users: write a function that does the work of the `detex` utility. It should take a string argument that represents a LaTeX file and strips the document preamble and any backslash commands. It should print its output to the terminal.

6.10 **Web**: The symbol `?` has another use in regular expressions that we have not discussed. Snoop around on the web, find out what it is, explain it, and write a short program showing how it can be used.

7 Text Manipulation

The previous chapter covered regular expressions and pattern matching in response to the fact that language researchers are often interested in sifting through texts and finding words or phrases with particular properties.

In this chapter, we focus on manipulating text, converting one string of letters into another. This is another task that comes up quite often when writing programs that deal with natural language.

We have, in fact, already done some of this in previous chapters. For example, in Section 6.4, we developed a program that calculated how often different consonant clusters occurred as word onsets in the *Alice* text. This program effectively translated from words to onsets using pattern matching and backreferences via the `match` object methods `start()` and `end()`.

In this chapter, we'll explain more general and powerful ways to do this using the functions `re.sub()`, `str.translate()`, `re.split()`, and the string method `join()`. We conclude the chapter by implementing a large example program that does a bit of English morphology.

7.1 String Manipulation Is Costly

One thing to keep in mind from the start is that string manipulation of any sort is computationally intensive. Consider a simple example like the following, where we go through a string letter by letter, stripping out vowels.

```
s = ''
for i in 'Apalachicola':
    if i not in 'aeiou':
        s = s + i
print(s)
```

Notice how every time we see a nonvowel, we invoke the line `s = s + 1`. What this does is create a new string every time it is invoked, save that new string as `s`, and make the old string available for garbage collection. If the string is long – or if you're doing this on a lot of strings – this will create a lot of new strings and call for a lot of garbage collection.

One can try to avoid this in various ways. For example, one might first convert the strings to lists, remove the vowels from the lists, and then convert the lists back to strings. While this avoids a certain amount of string processing, it does so by replacing it with conversions to and from lists.

Our point here is simply that string manipulation can be costly and should be done intelligently. For the small programs we present in this book, the cost is negligible, but in large programs, one needs to be careful.

7.2 Manipulating Text

The simplest function for manipulating text is `sub()` in the `re` module. This function converts one string into another by pattern matching: whatever part of the string matches the specified pattern is replaced with the specified replacement. The function takes those three arguments plus two more: a pattern, a replacement, the string, and a maximum number of replacements `count` plus additional flags `flags`. Here is a simple example:

```
import re

#define a string
s1 = 'This is a rather long string'
#replace '.s' with 'WOW'
s2 = re.sub('.s','WOW',s1)
#print old and new strings
print(s1,'\n',s2)
```

manip1.py

We can see the `count` variable at work in the next example:

```
import re

#a test string
s1 = 'This is a rather long string'
pat = '.s'       #a pattern
#find how many instances of the pattern
countmax = len(re.findall(pat,s1))
print(s1)        #print the string
i = 1            #make substitutions 1 by 1
while i < countmax+1:
```

```
    #make a change
    s2 = re.sub(pat,'WOW',s1,count=i)
    #print that one change
    print('\t',i,':',s2)
    i += 1          #increment counter
```

manip2.py

Here, we first identify how many instances of the pattern there are in the string with `findall()` and we then iterate through different numbers of substitutions. Note that when `count = 0`, we make all substitutions rather than none of them.

Finally, the `flags` argument provides several options. The most frequent one is case-insensitive matching: `flags=RE.i`. Here's an example:

```
import re

#a test string
s1 = 'This is a rather long string'
#do a replacement
s2 = re.sub('t','WOW',s1)
#do a case-insensitive replacement
s3 = re.sub('t','WOW',s1,flags=re.I)
#incorporate case directly in the pattern
s4 = re.sub('t|T','WOW',s1)
#show all three results
print(s1,'\n',s2,'\n',s3,'\n',s4,sep='')
```

manip3.py

Notice here how the effect of case-insensitive matching can be achieved by adjusting the pattern instead.

If your substitutions are all converting single letters to other single letters you can make use of the efficient `translate()` string method. You first make use of the `str.maketrans()` method to make a *translation table*, which specifies which letters are mapped to which; you then use that table to make the translation. Here's an example:

```
#make a translation table
mytab = str.maketrans('aeiou','happy')
#a test string
s = 'This is my sample string'
```

```
print(s)                    #print that string
print(s.translate(mytab)) #print translation
```

manip4.py

A translation table is implemented as a Python dictionary. Note that the two string arguments to `str.maketrans()` must be the same length.

Another function that is quite useful is `re.split()`. Recall that the string method `split()` splits a string up based on some specific delimiter string. The `re.split()` function allows you to split a string based on a regular expression instead. Here's an example:

```
import re

#a test string
s = 'First sentence. Second sentence.'
ss1 = s.split('e.')     #do a regular split
ss2 = re.split('e.',s) #do re.split
print(s)                #print sentence
#print split() results
print('s.split()')
for ss in ss1:
   print('\t"',ss,'"',sep='')
#print re.split() results
print('re.split()')
for ss in ss2:
   print('\t"',ss,'"',sep='')
```

manip5.py

Here we invoke the string method `split()` and `re.split()` with the string `'e.'`. The string method `split()` interprets this literally and splits the string into three strings; `re.split()` interprets this as a regular expression and splits the string into eight strings. Note, too, that the syntax for these is different. The `re.split()` function takes two arguments where the string to split is the second argument. The `split()` method is instead suffixed to the string it operates on and takes a single argument.

Finally, we have the string method `join()`, which joins a list of strings together with a string infix. The syntax is a bit unintuitive. The string it is suffixed to is the infix. It takes a single argument, which is a list of strings. Here's a simple example:

```
s = 'This is a sentence.'      #a test sentence
wds = s.split()                #split into words
hyphen = '-'                   #define hyphen
#join bits with hyphen
hyphenated = hyphen.join(wds)
#print original sentence
print(s)
#print hyphenated sentence
print(hyphenated)
```

manip6.py

7.3 Morphology

In this section, we build a stemming program for English. In particular, the program will remove suffixes so that words like *running*, *obfuscation*, *looks*, etc. become *run*, *obfuscate*, *look*, etc. The particular algorithm we will implement is a classic one: Porter (1980).[1]

Why do this? A stemming program like this can be viewed in several ways. One is to think of it as a theory of morphological decomposition: a model of how speakers break words up into meaningful units. Another way to think of a stemmer is simply as a practical tool so that we can find words that are morphologically related. For example, if we were searching for documents having to do with horses, it stands to reason that we would be interested in documents that contained the word *horse* or *horses*.

With Porter's algorithm, suffixes are removed from words in rule blocks. In the first block, one set of suffixes is removed; in the next block a different set of suffixes is removed, and so on. Each block is a set of rules that are organized disjunctively. Each rule in the block generally removes or rewrites one suffix. If more than one rule in a block is applicable, then the rule that is most specific applies and the others do not. For example, a block might contain these rules:

sses	→	ss
ies	→	i
ss	→	ss
s	→	∅

This would map *caresses* to *caress*, *carries* to *cari*, *looks* to *look*, etc. Note that the disjunctive requirement entails that *caresses* is not mapped to *caresse*, but to *caress*, because the first rule is more specific than the last.

[1] Porter, M. F. (1980) "An algorithm for suffix stripping," *Program* 14, 130–137.

Finally, rules are subject to conditions. Most often this is a condition on the size of the stem, but it can include other information as well, e.g., whether the final consonant is doubled or the identity of the final consonant.

We will take our task to be writing a function that takes a word and stems it. Once we've gotten this function in order, we can then apply it to words in *Alice* or, indeed, to anything else. Let's write some framing code that we can use to test our stemming function on a command-line argument:

```
import sys

def stem(w):            #stemming function frame
   return w

#get word from command-line
word = sys.argv[1]
root = stem(word)     #stem it!
#print word and its stem
print(word,':\t',root,sep='')
```

manip7.py

This code takes a command-line argument, applies the function `stem()` to it, and returns the output. To get a sense of what we're after, we might enrich `stem()` like this:

```
import re,sys

#stemming function for words in -ed
def stem(w):
   #does the word end in ed?
   m = re.search('(^.*)ed$',w)
   if m:                    #if it does...
      return m.group(1)     #return the stem
   else:                    #if it doesn't...
      return w              #just return word

#get word from command-line
word = sys.argv[1]
root = stem(word)           #stem it
#print word and stem
print(word,':\t',root,sep='')
```

manip8.py

This simply examines a word to see if it ends in *-ed* and, if so, removes it. Other words are unaffected. This is not a general solution to the stemming problem, but it shows the general logic of what we want to do.

The first step in our implementation is to code up Porter's *measure* function. Many of his rules depend on the size of the remaining stem when some putative suffix is removed. The measure of a stem is defined in terms of consonants and vowels. Specifically a consonant is defined as a letter other than *a*, *e*, *i*, *o*, *u*, or *y*. In the case of *y*, it is a consonant if it is not preceded by a consonant. We therefore start with a function to determine whether some specific letter in a string is a consonant.

```
import re,sys

#test if letter is consonant
#with respect to word!
def consonant(s,i):
   letter = s[i] #get relevant letter
   #it's not a consonant if it's aeiou
   if letter in 'aeiou':
      return False
   #word-initial y is a consonant
   elif letter == 'y' and i == 0:
      return True
   #it's a vowel if it follows a consonant
   elif letter == 'y' and consonant(s,i-1):
      return False
   else:          #otherwise it's a consonant
      return True

#stemming function frame again
def stem(w):
   return w

#get the command-line argument
word = sys.argv[1]
print(word)       #print it
#code to test the consonant() function
for i in range(len(word)):
   if consonant(word,i):
      print('C',end='')
   else:
```

```
        print('V',end='')
print()
```

manip9.py

We've added this to the program with the `stem()` function shell with framing code so we can test the function on strings given as a command-line argument.

If we take C to be a consonant and V to be a vowel, Porter treats all stems as matching this regular expression: `C*(V+C+)*V*`. The measure of a stem is defined as the number of times the `(V+C+)*` part matches. Another, simpler way to look at this is that the measure of a string is the number of times the sequence `VC` occurs. Porter gives these examples:

$m = 0$	tr, ee, tree, y, by
$m = 1$	trouble, oats, trees, ivy
$m = 2$	troubles, private, oaten, orrery

Our implementation of the measure function converts a stem to `C`s and `V`s and then returns the number of times the sequence `VC` occurs.

```
import re,sys

#checks if some element in
#a string is a consonant
def consonant(s,i):
   letter = s[i]
   if letter in 'aeiou':
      return False
   elif letter == 'y' and i == 0:
      return True
   elif letter == 'y' and consonant(s,i-1):
      return False
   else:
      return True

#converts string to Cs and Vs
def cv(w):
   res = ''
   for i in range(len(w)):
      if consonant(w,i):
         res += 'C'
```

```
        else:
            res += 'V'
    return res

#returns the measure of a string
def measure(w):
    cvword = cv(w)
    vcs = re.findall('VC',cvword)
    return len(vcs)

def stem(w):   #stemming code frame
    return w

word = sys.argv[1]
print(word)
print(cv(word))
print(measure(word))
```

manip10.py

The `measure()` function uses another function, `cv()`, which converts a string into Cs and Vs. We also use this function to display that intermediate output in the framing code in the program.

We're now ready to consider the rules of the system. Porter's rules have three main components. First, there is a test to see if the form ends in some particular ending string of letters. Second, there are several tests on the remainder that precedes that ending string. If those tests are true, the ending is mapped to a different ending. For example:

$(m > 1)$ ement $\rightarrow \emptyset$

Here, we first ask if the word ends in the string *ement*. Second, we ask if the measure of the remaining material is greater than 1. If both of those hold, we map *ement* to null. This rule would fail to apply to *happiness* because it does not end in *ement*. This rule would fail to apply to *cement* because the measure of *c* is 0. Finally, this rule would apply to *requirement* because the measure of *requir* is 2.

Porter's conditions also include things like the following:

∗S	–	the stem ends with *s* (and similarly for the other letters).
∗V∗	–	the stem contains a vowel.
∗D	–	the stem ends with a double consonant (e.g., *tt*, *ss*).
∗O	–	the stem ends CVC, where the second C is not *w*, *x*, or *y* (e.g., *-wil*, *-hop*).

These are expressed above in a version of Porter's notation; we will convert these to regular expressions. The conditions on a rule can be more complex: multiple separate conditions joined together with *and* or *or*. For example:

(m > 1 and (∗s or ∗t))

This condition tests for a stem where $m > 1$ and that ends in *s* or *t*. Here's another:

(∗D and not (∗L or ∗S or ∗Z))

This tests for a stem ending with a double consonant other than *l*, *s*, or *z*.

Let's now add *general* code for handling such conditions to our system. Here is a first pass:

```
import re,sys

#is some element in string a consonant?
def consonant(s,i):
   letter = s[i]
   if letter in 'aeiou':
      return False
   elif letter == 'y' and i == 0:
      return True
   elif letter == 'y' and consonant(s,i-1):
      return False
   else:
      return True

#converts string to Cs and Vs
def cv(w):
   res = ''
   for i in range(len(w)):
      if consonant(w,i):
         res += 'C'
      else:
         res += 'V'
   return res

#returns measure of a string
def measure(w):
   cvword = cv(w)
   vcs = re.findall('VC',cvword)
   return len(vcs)
```

```
#general rule framework
def rule(c,e,r,w):
   m = re.search('^(.*)'+e+'$',w)
   if m:
      s = m.group(1)
      if c(s):
         return s+r
   return None

def m1cond(x):    #condition: m > 0
   if measure(x) > 0:
      return True
   return False

#specific sample rule for -ed
def edrule(w):
   x = rule(m1cond, 'ed', '', w)
   return x

def stem(w):      #using -ed rule
   res = edrule(w)
   if res:
      return res
   return w

word = sys.argv[1]
print(word)
print(stem(word))
```

manip11.py

There are four basic changes here. First, we've added a general form for rules called `rule()`. This function takes four arguments: `c`: a condition, `e`: the suffix, `r`: what the suffix is replaced with, and `w`: the word we are applying the rule to. The idea is that any specific rule we want can be defined in terms of `rule()`.

We then use this to define a sample rule that we call `edrule()`. This is a simple example of what a rule might look like. This rule rewrites *-ed* as null just in case the measure of the remaining stem is greater than 0. The condition on the rule is formalized as `m1cond()`.

Finally, we add the function `edrule()` to our `stem()` function. If the word satisfies `edrule()`, we return that result. If not, we just return the word.

All of this is called with a command-line argument so we can play around with different word types.

We've got a fair amount of code now, so let's separate the general code off into a callable module. The idea would be that we could build different stemmers that could draw on this code. Here's the module:

```
import re,sys

#is some element in string a consonant?
def consonant(s,i):
   letter = s[i]
   if letter in 'aeiou':
      return False
   elif letter == 'y' and i == 0:
      return True
   elif letter == 'y' and consonant(s,i-1):
      return False
   else:
      return True

def cv(w):        #convert string to Cs and Vs
   res = ''
   for i in range(len(w)):
      if consonant(w,i):
         res += 'C'
      else:
         res += 'V'
   return res

def measure(w): #returns measure of a string
   cvword = cv(w)
   vcs = re.findall('VC',cvword)
   return len(vcs)

#general rule framework
def rule(c,e,r,w):
   m = re.search('^(.*)'+e+'$',w)
   if m:
      s = m.group(1)
      if c(s):
         return s+r
   return None
```

manip12.py

We then call it with something like this:

```
import re,sys,manip12

def m1cond(x):    #condition: m > 0
   if manip12.measure(x) > 0:
      return True
   return False

def edrule(w):    #sample rule for -ed
   x = manip12.rule(m1cond, 'ed', '', w)
   return x

def stem(w):      #stemming with -ed rule
   res = edrule(w)
   if res:
      return res
   return w

word = sys.argv[1]
print(word)
print(stem(word))
```

manip13.py

As discussed, Porter's algorithm divides rules up into eight blocks:

Step	1	a
		b
		c
Step	2	
Step	3	
Step	4	
Step	5	a
		b

Let's elaborate our system a bit to anticipate this:

```
import re,sys,manip12

def m1cond(x):    #condition: m > 0
   if manip12.measure(x) > 0:
      return True
   return False
```

```
def step1a(w):
   return w

def step1b(w):
   return w

def step1c(w):
   return w

def step2(w):
   return w

def step3(w):
   return w

def step4(w):
   return w

def step5a(w):
   return w

def step5b(w):
   return w

def stem(w):        #stemming with blocks
   s1a = step1a(w)
   s1b = step1b(s1a)
   s1c = step1c(s1b)
   s2 = step2(s1c)
   s3 = step3(s2)
   s4 = step4(s3)
   s5a = step5a(s4)
   s5b = step5b(s5a)
   return s5b

word = sys.argv[1]
print(word)
print(stem(word))
```

manip14.py

Here we've simply indicated that the function `stem()` is a set of blocks that apply in sequence. We've then added placeholders for each of those blocks.

Porter's first block – step 1a – is the one we gave above on page 142. We implement this in the function `step1a()`.

```
import re,sys
from manip12 import *

def m1cond(x):     #condition: m > 0
   if measure(x) > 0:
      return True
   return False

def nullcond(x):  #a vacuous condition
   return True

def step1a(w):
   a = rule(nullcond,'sses','ss',w)
   if a: return a
   b = rule(nullcond,'ies','i',w)
   if b: return b
   c = rule(nullcond,'ss','ss',w)
   if c: return c
   d = rule(nullcond,'s','',w)
   if d: return d
   return w

def step1b(w):
   return w

def step1c(w):
   return w

def step2(w):
   return w

def step3(w):
   return w

def step4(w):
   return w

def step5a(w):
   return w
```

```
def step5b(w):
   return w

def stem(w):          #stemming with blocks
   s1a = step1a(w)
   s1b = step1b(s1a)
   s1c = step1c(s1b)
   s2 = step2(s1c)
   s3 = step3(s2)
   s4 = step4(s3)
   s5a = step5a(s4)
   s5b = step5b(s5a)
   return s5b

word = sys.argv[1]
print(word)
print(stem(word))
```

manip15.py

There are a couple of things to note here. First, we import from `manip12.py` a little differently so we can use its functions without the `manip12.` prefix. Second, our rule format requires a condition, so we've implemented a vacuous condition `nullcond()` to accommodate cases where Porter's rules have no condition. Finally, we've implemented the disjunctive property of Porter's rules by checking if a rule has applied. If it does, we immediately exit the block with a `return`.

Let's now go on to the next block Step 1b:

Step 1b

Condition	Change	Example
m > 0	eed → ee	feed → feed
		agreed → agree
∗V∗	ed → ∅	plastered → plaster
		bled → bled
∗V∗	ing → ∅	motoring → motor
		sing → sing

Remember that these three rules apply disjunctively. Here is a first pass at implementing this block:

```
import re,sys
from manip12 import *

def m1cond(x):     #condition: m > 0
```

```
    if measure(x) > 0:
        return True
    return False

def nullcond(x):    #a vacuous condition
    return True

def vcond(x):       #condition: contains vowel
    cvform = cv(x)
    if re.search('V',cvform):
        return True
    return False

def step1a(w):
    a = rule(nullcond,'sses','ss',w)
    if a: return a
    b = rule(nullcond,'ies','i',w)
    if b: return b
    c = rule(nullcond,'ss','ss',w)
    if c: return c
    d = rule(nullcond,'s','',w)
    if d: return d
    return w

def step1b(w):
    a = rule(m1cond,'eed','ee',w)
    if a: return a
    b = rule(vcond,'ed','',w)
    if b: return b
    c = rule(vcond,'ing','',w)
    if c: return c
    return w

def step1c(w):
    return w

def step2(w):
    return w

def step3(w):
    return w
```

```
def step4(w):
   return w

def step5a(w):
   return w

def step5b(w):
   return w

def stem(w):         #stemming with blocks
   s1a = step1a(w)
   s1b = step1b(s1a)
   s1c = step1c(s1b)
   s2 = step2(s1c)
   s3 = step3(s2)
   s4 = step4(s3)
   s5a = step5a(s4)
   s5b = step5b(s5a)
   return s5b

word = sys.argv[1]
print(word)
print(stem(word))
```

manip16.py

The `step1b()` function makes use of the `m1cond()` condition and a new condition `vcond()`. There is another wrinkle, however. If either of the *-ed* or *-ing* rules applies, the following special block applies:

Special block

Condition	Change	Example
	at → ate	conflat(ed) → conflate
	bl → ble	troubl(ed) → trouble
	iz → ize	siz(ed) → size
(*D and not (*l or *s or *z))	. . . → single letter	hopp(ing) → hop
		tann(ed) → tan
		fall(ing) → fall
		hiss(ing) → hiss
		fizz(ed) → fizz
(m = 1 and *O)	Ø → E	fail(ing) → fail
		fil(ing) → file

The fourth rule here is rather complex both in terms of the condition on its application and in terms of what it does, so we will write a special rule for this. First, we add a new special block and have it apply just in case the *-ed* or *-ing* rules apply.

```
import re,sys
from manip12 import *

def m1cond(x):      #condition: m > 0
   if measure(x) > 0:
      return True
   return False

def nullcond(x):    #a vacuous condition
   return True

def vcond(x):       #condition: contains vowel
   cvform = cv(x)
   if re.search('V',cvform):
      return True
   return False

def step1a(w):
   a = rule(nullcond,'sses','ss',w)
   if a: return a
   b = rule(nullcond,'ies','i',w)
   if b: return b
   c = rule(nullcond,'ss','ss',w)
   if c: return c
   d = rule(nullcond,'s','',w)
   if d: return d
   return w

def special(w):     #special block for ed/ing
   return w

def step1b(w):
   a = rule(m1cond,'eed','ee',w)
   if a: return a
   b = rule(vcond,'ed','',w)
   if b: return special(b)
   c = rule(vcond,'ing','',w)
```

```
    if c: return special(c)
    return w

def step1c(w):
    return w

def step2(w):
    return w

def step3(w):
    return w

def step4(w):
    return w

def step5a(w):
    return w

def step5b(w):
    return w

def stem(w):        #stemming with blocks
    s1a = step1a(w)
    s1b = step1b(s1a)
    s1c = step1c(s1b)
    s2 = step2(s1c)
    s3 = step3(s2)
    s4 = step4(s3)
    s5a = step5a(s4)
    s5b = step5b(s5a)
    return s5b

word = sys.argv[1]
print(word)
print(stem(word))
```

manip17.py

We now flesh out the contents of the `special()` block.

```
import re,sys
from manip12 import *
```

```
def m1cond(x):      #condition: m > 0
   if measure(x) > 0:
      return True
   return False

def nullcond(x):    #vacuous condition
   return True

def vcond(x):       #condition: contains vowel
   cvform = cv(x)
   if re.search('V',cvform):
      return True
   return False

def step1a(w):
   a = rule(nullcond,'sses','ss',w)
   if a: return a
   b = rule(nullcond,'ies','i',w)
   if b: return b
   c = rule(nullcond,'ss','ss',w)
   if c: return c
   d = rule(nullcond,'s','',w)
   if d: return d
   return w

#specialrule
def specialrule(w):
   cvform = cv(w)
   m = re.search('CC$',cvform)
   if m:
      m2 = re.search('^(.*)([^szl])\\2',w)
      if m2:
         return m2.group(1)+m2.group(2)
   return None

def m1ocond(w):     #m1ocond
   cvform = cv(w)
   if measure(cvform) == 1:
      m = re.search('CVC$',cvform)
      if m:
         m2 = re.search('[^xyw]$',w)
         if m2:
```

```
            return True
    return False

def special(w):    #special block for ed/ing
    a = rule(nullcond,'at','ate',w)
    if a: return a
    b = rule(nullcond,'bl','ble',w)
    if b: return b
    c = rule(nullcond,'iz','ize',w)
    if c: return c
    d = specialrule(w)
    if d: return d
    e = rule(m1ocond,'','e',w)
    if e: return e
    return w

def step1b(w):
    a = rule(m1cond,'eed','ee',w)
    if a: return a
    b = rule(vcond,'ed','',w)
    if b: return special(b)
    c = rule(vcond,'ing','',w)
    if c: return special(c)
    return w

def step1c(w):
    return w

def step2(w):
    return w

def step3(w):
    return w

def step4(w):
    return w

def step5a(w):
    return w

def step5b(w):
    return w
```

```
def stem(w):        #stemming with blocks
   s1a = step1a(w)
   s1b = step1b(s1a)
   s1c = step1c(s1b)
   s2 = step2(s1c)
   s3 = step3(s2)
   s4 = step4(s3)
   s5a = step5a(s4)
   s5b = step5b(s5a)
   return s5b

word = sys.argv[1]
print(word)
print(stem(word))
```

manip18.py

There are a couple of complications in the `special()` block. First, we must implement a new condition `m1ocond()`. Second, the rule that eliminates final double letters doesn't really follow our `rule()` format, so we craft a special rule for it with the unimaginative name `specialrule()`. That rule is interesting because of the complexity of the regular expression it uses. We need to identify stems where the final letter is doubled, and then we need to be able to return everything but the final letter. We use a backreference in the pattern to do this. Note that when a backreference is used in a pattern like this, we must use an extra backslash: `'\\2'`.

There are a number of remaining steps to implement, but they are straightforward given what we have done so far. The next version of the code is the complete stemming algorithm with all these additional steps filled in.

```
import re,sys
from manip12 import *

def m1cond(x):      #condition: m > 0
   if measure(x) > 0:
      return True
   return False

def m2cond(x):      #condition: m > 1
   if measure(x) > 1:
      return True
   return False
```

```
def nullcond(x):   #vacuous condition
   return True

def vcond(x):      #condition: contains vowel
   cvform = cv(x)
   if re.search('V',cvform):
      return True
   return False

def step1a(w):
   a = rule(nullcond,'sses','ss',w)
   if a: return a
   b = rule(nullcond,'ies','i',w)
   if b: return b
   c = rule(nullcond,'ss','ss',w)
   if c: return c
   d = rule(nullcond,'s','',w)
   if d: return d
   return w

#specialrule
def specialrule(w):
   cvform = cv(w)
   m = re.search('CC$',cvform)
   if m:
      m2 = re.search('^(.*)([^szl])\\2',w)
      if m2:
         return m2.group(1)+m2.group(2)
   return None

def m1ocond(w):    #m=1 and ends in CV[^xyw]
   cvform = cv(w)
   if measure(cvform) == 1:
      m = re.search('CVC$',cvform)
      if m:
         m2 = re.search('[^xyw]$',w)
         if m2:
            return True
   return False

def special(w):    #special block for ed/ing
```

```
    a = rule(nullcond,'at','ate',w)
    if a: return a
    b = rule(nullcond,'bl','ble',w)
    if b: return b
    c = rule(nullcond,'iz','ize',w)
    if c: return c
    d = specialrule(w)
    if d: return d
    e = rule(m1ocond,'','e',w)
    if e: return e
    return w

def step1b(w):
    a = rule(m1cond,'eed','ee',w)
    if a: return a
    b = rule(vcond,'ed','',w)
    if b: return special(b)
    c = rule(vcond,'ing','',w)
    if c: return special(c)
    return w

def step1c(w):
    a = rule(vcond,'y','i',w)
    if a: return a
    return w

def step2(w):
    a = rule(m1cond,'ational','ate',w)
    if a: return a
    b = rule(m1cond,'tional','tion',w)
    if b: return b
    c = rule(m1cond,'enci','ence',w)
    if c: return c
    d = rule(m1cond,'anci','ance',w)
    if d: return d
    e = rule(m1cond,'izer','ize',w)
    if e: return e
    f = rule(m1cond,'abli','able',w)
    if f: return f
    g = rule(m1cond,'alli','al',w)
    if g: return g
    h = rule(m1cond,'entli','ent',w)
```

```
    if h: return h
    i = rule(m1cond,'eli','e',w)
    if i: return i
    j = rule(m1cond,'ousli','ous',w)
    if j: return j
    k = rule(m1cond,'ization','ize',w)
    if k: return k
    l = rule(m1cond,'ation','ate',w)
    if l: return l
    m = rule(m1cond,'ator','ate',w)
    if m: return m
    n = rule(m1cond,'alism','al',w)
    if n: return n
    o = rule(m1cond,'iveness','ive',w)
    if o: return o
    p = rule(m1cond,'fulness','ful',w)
    if p: return p
    q = rule(m1cond,'ousness','ous',w)
    if q: return q
    r = rule(m1cond,'aliti','al',w)
    if r: return r
    s = rule(m1cond,'iviti','ive',w)
    if s: return s
    t = rule(m1cond,'biliti','ble',w)
    if t: return t
    return w

def step3(w):
    a = rule(m1cond,'icate','ic',w)
    if a: return a
    b = rule(m1cond,'ative','',w)
    if b: return b
    c = rule(m1cond,'alize','al',w)
    if c: return c
    d = rule(m1cond,'iciti','ic',w)
    if d: return d
    e = rule(m1cond,'ical','ic',w)
    if e: return e
    f = rule(m1cond,'ful','',w)
    if f: return f
    g = rule(m1cond,'ness','',w)
    if g: return g
```

```
    return w

def m2stcond(w):   #m > 1 and ends in [st]
    if m2cond(w):
        m = re.search('[st]$',w)
        if m:
            return True
    return False

def step4(w):
    a = rule(m2cond,'al','',w)
    if a: return a
    b = rule(m2cond,'ance','',w)
    if b: return b
    c = rule(m2cond,'ence','',w)
    if c: return c
    d = rule(m2cond,'er','',w)
    if d: return d
    e = rule(m2cond,'ic','',w)
    if e: return e
    f = rule(m2cond,'able','',w)
    if f: return f
    g = rule(m2cond,'ible','',w)
    if g: return g
    h = rule(m2cond,'ant','',w)
    if h: return h
    i = rule(m2cond,'ement','',w)
    if i: return i
    j = rule(m2cond,'ment','',w)
    if j: return j
    k = rule(m2cond,'ent','',w)
    if k: return k
    l = rule(m2stcond,'tion','',w)
    if l: return l
    m = rule(m2cond,'ou','',w)
    if m: return m
    n = rule(m2cond,'ism','',w)
    if n: return n
    o = rule(m2cond,'ate','',w)
    if o: return o
    p = rule(m2cond,'iti','',w)
    if p: return p
```

```
    q = rule(m2cond,'ous','',w)
    if q: return q
    r = rule(m2cond,'ive','',w)
    if r: return r
    s = rule(m2cond,'ize','',w)
    if s: return s
    return w

#m = 1 and not *O
def m1notocond(w):
    if measure(w) != 1:
        return False
    cvform = cv(w)
    m = re.search('CVC$',cvform)
    if m:
        m2 = re.search('[wxy]$',w)
        if m2:
            return False
    return True

def step5a(w):
    a = rule(m2cond,'e','',w)
    if a: return a
    b = rule(m1notocond,'e','',w)
    if b: return b
    return w

def step5b(w):
    if m2cond(w):
        m = re.search('(^.*l)(l)$',w)
        if m:
            return m.group(1)
    return w

def stem(w):        #stemming with blocks
    s1a = step1a(w)
    s1b = step1b(s1a)
    s1c = step1c(s1b)
    s2 = step2(s1c)
    s3 = step3(s2)
    s4 = step4(s3)
    s5a = step5a(s4)
```

```
        s5b = step5b(s5a)
        return s5b

    word = sys.argv[1]
    print(word)
    print(stem(word))
```

manip19.py

This is a fairly large piece of code, but we have tried to modularize it in several ways. First, we have factored out generic code and put it into a separate module `manip12.py`. Second, we have posited a general rule format and put that in `manip12.py` as well. Finally, following Porter's own description of the algorithm, we have broken it into steps and made each of these its own function.

7.4 Exercises

7.1 Imagine you have some sentence `s` and you do this:

```
' '.join(s.split())
```

Will that have any effect? If so, what and when?

7.2 The stemming algorithm in the previous section suffers from the absence of a general technique for disjunction: we have to keep repeating `if ...: return ...`. Can you think of a way to avoid this?

7.3 The code for the stemming algorithm is repetitive in that it calls `rule()` multiple times in each step. Revise the code to avoid this.

7.4 Why would the `sub()` function be harder to use in our stemmer?

7.5 Write a program that strips off English prefixes.

7.6 Write a program that plays Pig Latin.

7.7 Use the `translate()` method to create a function that does the work of `upper()`.

7.8 Write your own function that will do the work of `translate()`.

7.9 Write a program that generates verbal paradigms for an inflectional language you know. For example, your program might generate the present tense paradigm for verbs in *-ar* in Spanish.

7.10 Write a program that converts active sentences in English to passive sentences. This means changing the verbal morphology *and* adjusting word order as required.

7.11 **Web**: There is also a function `re.subn()` that we have not discussed. Snoop around to find out what it does, explain it, and write a program that makes good use of it.

8 Internet Data

In this chapter, we discuss how to obtain and handle data from the internet. This is a huge topic, so we can only scratch the surface here.

Specifically, we talk about how to retrieve webpages and extract text or other information from them. This entails a discussion of the structure of HTML documents and methods for getting various sorts of information out of them. It also requires we deal with different text encodings.

Working with web data quickly leads to issues of efficiency. Retrieving any web page requires interacting with other computers over the web. This means that your program may have to wait for other systems to respond. We therefore introduce some simple methods for parallelizing your code so that these interactions can be as efficient as possible.[1]

We then turn to text encodings. Text can be encoded in a web page – and other documents – in many ways. We must therefore understand those encodings and know how to work with them.

Finally, we conclude the chapter with a program for a simple webcrawler, a program that starts from a single webpage and then recursively follows and retrieves links with particular properties.

8.1 Retrieving Webpages

Retrieving webpages is extremely simple. The `urllib.request` module includes a function, `urlopen()`, which creates a stream that can be read from. The following program exemplifies:

```
import urllib.request

#a url to read from
link = "http://www.u.arizona.edu/~hammond/"
#open a link to the url
f = urllib.request.urlopen(link)
```

[1] We deal with parallelism more fully in Chapter 11.

```
#read the page
myfile = f.read()
#print the decoded page
print(myfile.decode('UTF-8'))
```

web1.py

Here we import the relevant module. We open a connection with `urlopen()`, and then read from it with `read()`. We convert that to a readable text format with the string method `decode('UTF-8')`; we cover this in depth in Section 8.5 below. We can then print out what is read.

Note that a simple webpage, like the one we read here, is actually text intermixed with various sorts of formatting commands. If we want to make sense of what we read in, we need to know a bit about that formatting.

8.2 HTML

When you surf the web, you are using your browser to retrieve data from other computers. That data can be in a variety of formats. For example, you can retrieve pictures, music, movies, pdf documents, etc.

You most commonly see text documents formatted using HTML or *hypertext markup language*. Following is an extremely simple example of what such a document might look like behind the scenes:

```
<!DOCTYPE html>
<html>
   <head>
      <title>a title</title>
   </head>
   <body>
      <h1>a heading</h1>
      <p>content</p>
   </body>
</html>
```

Such a document, however, doesn't display like that in your web browser. Rather, all the things in angled brackets are invisible instructions for how your browser should display what. The page just shown might display like this:

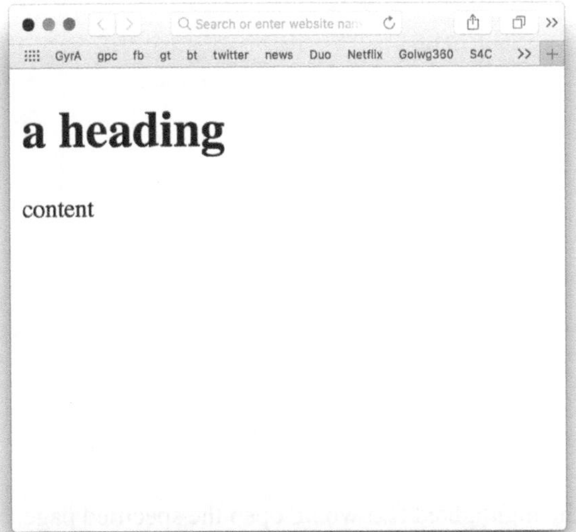

HTML is built around *tags*, instructions to your web browser that are marked with angled brackets. For example, `<p>` marks a paragraph, `</em>` marks the end of a span of emphasized text, etc.

Those tags may or may not come in pairs. For example, in the example above, the document declaration at the beginning that indicates that the document is written in HTML (version 5) is unpaired. On the other hand, the tags marking the *head* of the document `<head>` and `</head>` are paired. When tags are paired, they typically have that structure; the opening tag is marked simply with angled brackets, and the closing tag has angled brackets and a slash after the left bracket. In the example above, we've indented lines to show this structure more clearly, but unlike in Python, this indentation is not required.

Tags can also have attributes, additional information specified in the tag. For example, a hyperlink, text that you can click to go to a different web page, has this structure:

```
Here is a link.
<a href="http://www.google.com">Go
to Google.</a>
That was a link.
```

Here there is a paired tag `a` with an opening and a closing tag. The opening tag has an attribute `href`, which is specified for a web address. This might display like this:

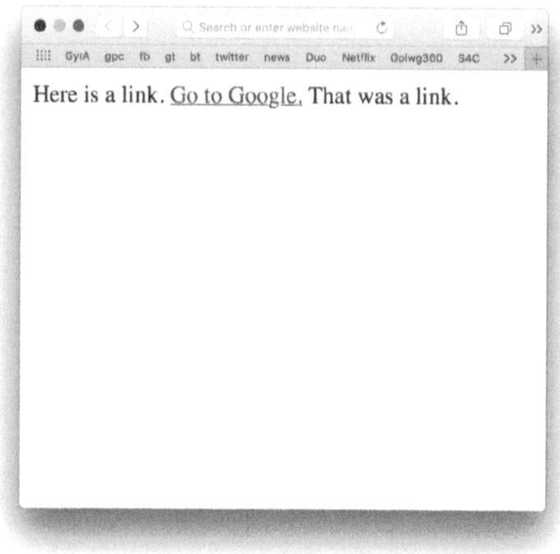

Clicking on the highlighted text would open the specified page.

The general structure of a page marked up with HTML is otherwise a little underdetermined – a little wild west. There are two reasons for this. First, there are a number of versions of HTML in use on the web, and they do not all have the same requirements. Second, individual web browsers sometimes allow for browser-specific tags, or their own special interpretation of tags in general use. Our remarks here then should be taken as strongly qualified.

As exemplified above, an HTML document can begin with a document declaration. It may then delimit the entire remaining document with the tags <html>...</html>. The document is typically further broken up into two parts: the head and the body, which are marked with <head>...</head> and <body>...</body> respectively. The head of the document includes metadata about the document, the title, javascript code, color and formatting information, etc. The body of the document includes the text with various sorts of markup to structure the text or format spans of text.

We can't cover all the formatting possibilities, but here are some of the main ones.

Headings Documents can include formatted section headings at different levels, <h1>, <h2>, <h3>, etc. These occur in pairs: <h1>...</h1>.

Paragraphs Paragraphs are marked with <p>...</p>. Here the closing tag is *not* required. Line breaks can be marked with the unpaired tag <br>.

Lists There are various sorts of lists: ordered lists <ol>...</ol> and unordered lists <ul>...</ul>. List items are marked with <li>...</li>. The list item tag need not be paired.

Formatting Text can be formatted in several ways. There is specific formatting that says precisely how the text is to be formatted, e.g., italic <i>...</i> or boldface <b>...</b>. There is also logical formatting where the code leaves it up to the browser how to display the text, e.g., <strong>...</strong> or <em>...</em> (emphatic).

Here is a simple HTML document that includes many of these formats:

```
<!DOCTYPE html>
<html>
 <head>
  <title>Here's a title</title>
 </head>
 <body>
  <h1>Some examples</h1>
  <h2>Text chunks</h2>
  <p>This is a paragraph.</p>
  <p>This is another paragraph.<br>
     with a break in the middle.</p>
  <h2>Lists and formatting</h2>
  <ol>
   <li>Some <strong>strong</strong> text.</li>
   <li>Some <b>bold</b> text.</li>
   <li>Some <em>emphasized</em> text.</li>
   <li>Some <i>italic</i> text.</li>
  </ol>
 </body>
</html>
```

In my own browser, the preceding code displays as follows:

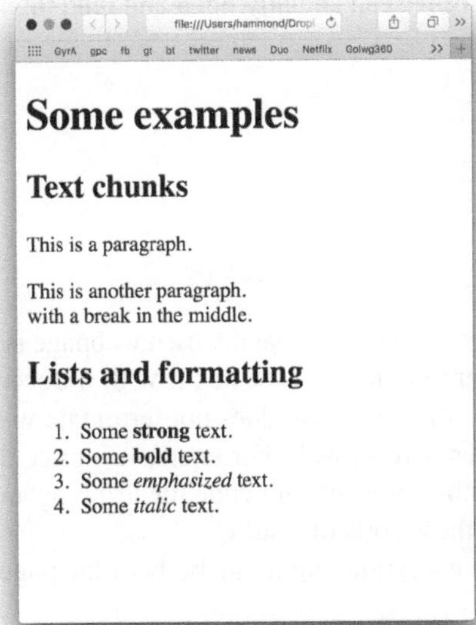

Notice how the settings for my browser render strong text as bold and emphasized text as italic. Your browser may display these differently.

Finally, web pages can be *dynamic*, rendered differently depending on day, time, location or browser of the user, etc. On the host side, this is often done with programming languages like `php`, but since this is handled on the hosting machine, you won't see this code. On the other hand, dynamic content can be rendered with `javascript` on the client side, which means that you can often find `javascript` code in webpages you download.

8.3 Parsing HTML

If you collect data via the web, you will most often be collecting data from text pages marked up with HTML. It is then necessary to remove or translate some or all of the markup.

How you do that depends tremendously on your goals. If all you want is the words on the page and you don't care about formatting or discourse or syntactic breaks, simply stripping the HTML may suffice. On the other hand, if you are interested in the larger structure of the webpage – where the section breaks are, the difference between line breaks, sentence breaks, and paragraph breaks – then you can't simply remove the HTML.

Let's assume, for convenience, that you want to remove the HTML. A priori, the simplest thing to do is to write code yourself to remove it. The code files for the text include my own webpage downloaded as `hammond.html`. For convenience, we will operate with this local file.

First, we make sure we can actually open and read the file:

```
f = open('hammond.html','r')
t = f.read()
f.close()
print(t)
```

web5.py

If you run this program, you'll see that my own webpage is not exactly in orthodox format. For example, it does not begin with a document declaration. In addition, the head of the document does not terminate with `</head>`. There are many other issues here as well. For example, notice that some of the tags are capitalized and others are not. Our code for stripping and translating HTML must accommodate these sorts of issues.

Let's first remove everything up to the body of the page.

```
import re

#open local file
f = open('hammond.html','r')
t = f.read()        #read whole thing in
f.close()           #close the stream
#do a multi-line substitution, deleting
#everything up to the body of the page
t = re.sub('^.*<body>','',t,flags=re.I|re.S)
print(t)            #print the result
```

web6.py

Here we are matching `.` against any character including a line return, so we must use the flag `re.S`. We are also anticipating that some tags are capitalized and some not, so we use `re.I` for that. As already noted, when we use both flags, they must be joined with `|`.

To remove all remaining tags, we again use `re.sub()`.

```
import re

#open local page
f = open('hammond.html','r')
t = f.read()        #read it all in
f.close()           #close file stream
#eliminate header up to body of page
t = re.sub('^.*<body>','',t,flags=re.I|re.S)
#remove all tags
t = re.sub('<[^>]*>',' ',t,flags=re.I|re.S)
print(t)            #print what's left
```

web7.py

Here we replace tags with spaces, which generally works, but running the program shows that this sometimes puts spaces where they shouldn't be. If we were removing punctuation later, this would not be a problem.

Another issue with this code is that it seems to leave the string `-->` scattered about the page. This is the right side of an HTML comment. Comments in HTML are of the form `<!--...-->` and it looks like the page source includes things like `<!--...</a>...-->`, where an HTML tag occurs inside a comment, and which our code above parses incorrectly. The following code gets at least some of these.

```
import re

#open local webpage
f = open('hammond.html','r')
t = f.read()        #read it all in
f.close()           #close file stream
#get rid of header
t = re.sub('^.*<body>','',t,flags=re.I|re.S)
#get rid of (at least some) html comments
t = re.sub(
    '<!--[^-]*-->',' ',t,flags=re.I|re.S
)
#get rid of at least some tags
t = re.sub('<[^>]*>',' ',t,flags=re.I|re.S)
print(t)             #print what remains
```

web8.py

This works better, but still doesn't get all cases.

We can keep tweaking this code to do better, but the lesson should be clear: it is hard to accurately and exactly translate or strip HTML code. If you want to write the code to do that yourself, you must pay special attention to why you are stripping HTML, so that you can focus your efforts on removing or translating what really needs to be removed or translated. You must also ultimately be prepared for a certain amount of noise in your data.

There is another way to go here, though. There are existing modules for parsing HTML. These are also not perfect, but they can often be a lot closer to perfect that you or I might have patience for!

One very common solution in Python is *Beautiful Soup*,[2] a free open-source module for Python for parsing and manipulating HTML. You have to install it yourself, but once installed, it can be called like any other module. The module is called `bs4` and the relevant function is `BeautifulSoup()`. What that does is parse the HTML and build a *document model*. This is a treelike representation of the HTML document that you can extract elements from easily. The following code exemplifies:

```
#import for reading urls
import urllib.request
#import for parsing html
from bs4 import BeautifulSoup
```

[2] www.crummy.com/software/BeautifulSoup/

```
#non-local page this time
link = "http://www.u.arizona.edu/~hammond/"
#connect to that page
f = urllib.request.urlopen(link)
myfile = f.read()        #read it all in
#build a document model
soup = BeautifulSoup(myfile,'html.parser')
print(myfile)            #print page verbatim
print(soup.prettify())   #pretty-print html
print(soup.get_text())   #extract the text
#got through all the hyperlinks...
for link in soup.find_all('a'):
   #...and print them
   print(link.get('href'))
```

web9.py

Here we read in a web page and then parse it with `BeautifulSoup()`. We can then print a pretty version of it with `prettify()`, extract the text with `get_text()`, or find all instances of a tag with `find_all()`. Each tag found is its own treelike representation, so we can continue to call methods on them. In the example at hand, we call the `get()` method to extract the text of the `href` attribute for the `a` tags.

Notice incidentally that the `get_text()` method does a better job of dealing with HTML comments in this example than the code we wrote above.

8.4 Parallelism

Once you start using web data as input in your program, you quickly run into a problem. When your program requests a web page, it relies on the speed of the network and the other computer you are requesting the page from. If you are getting a lot of web pages, this can really slow down your program. Here's an example of a program that shows this effect:

```
#import for timing your code
import time
#import for reading webpages
from urllib.request import urlopen

def mytime():    #return time in milliseconds
   return round(time.time() * 1000)
```

```
def myget(url): #read url and time that read
   start = mytime()
   data = urlopen(url,timeout=5).read()[:50]
   result = {"url": url, "data": data}
   now = str(mytime() - start)
   print(url + ": " + now + "ms")
   return result

#a random list of urls
urls = ['http://www.google.com/',
'http://www.yahoo.com/',
'http://golwg360.cymru/newyddion',
'https://news.google.com/news',
'https://tartarus.org/martin/PorterStemmer/',
'https://en.wikipedia.org/wiki/Main_Page',
'http://www.u.arizona.edu']

#start overall timing
start = mytime()
results = []    #list to collect results
#go through urls 1 by 1
for i in range(len(urls)):
   #get url and text read
   result = myget(urls[i])
   #append those to results
   results.append(result)
#get end time
now = str(mytime() - start)
#print overall time
print("Total = " + now + " ms\n")
```

web10.py

This program imports from the `urllib.request` module to get access to the `urlopen()` function. It also uses the `time()` function from the `time` module, which returns the current time (calculated from a specific and irrelevant time of origin). We can use the `time()` function to determine how long it takes our code to run. First, we define a function `mytime()`, which gets the current time in milliseconds. We also define a function `myget()`, which takes a url as an argument, reads the first 50 characters of that file, and then prints out how long it took to get that done. Note how the `urlopen()` function takes an additional argument that specifies how long it will wait for a webpage before moving on. We then define a fairly random list of urls and run through them with `myget()`. We also return the total amount of time for the whole program to run.

If you run this program, you'll see that the total run time is roughly the sum of the time it takes to retrieve each page. This is perhaps unsurprising. Notice, though, that most of the time your computer is simply waiting for the webpages to be delivered. That is, for each request, your program sends the request and then sits idle while it waits for the other machine to send its response.

A more efficient system would have your computer do something else while waiting for the other computer to respond.

A related issue is that most modern computers have a multiprocessor architecture, which means that at the processing level, setting aside hardware bottlenecks, they can do more than one thing at once.

We don't want to get into the nitty gritty of processes, threads, parallelism, etc., but we can do substantially better if we take advantage of these ideas.[3] The basic logic is that, with the `multiprocessing` module, you can do more than one thing at once, depending on your own computer hardware.

The following bit of code shows this:

```
import time        #for timing info
#to read webpages
from urllib.request import urlopen
#to do more than one thing at once
from multiprocessing import Pool

#current time in milliseconds
def mytime():
   return round(time.time() * 1000)

def myget(url):  #50 characters of a webpage
   start = mytime()
   data = urlopen(url,timeout=5).read()[:50]
   result = {"url": url, "data": data}
   now = str(mytime() - start)
   print(url + ": " + now + "ms")
   return result

#some random urls
urls = ['http://www.google.com/',
```

[3] We return to this topic in Chapter 11.

```
    'http://www.yahoo.com/',
    'http://golwg360.cymru/newyddion',
    'https://news.google.com/news',
    'https://tartarus.org/martin/PorterStemmer/',
    'https://en.wikipedia.org/wiki/Main_Page',
    'http://www.u.arizona.edu']

#print urls in order accessed
for i in range(len(urls)):
    print(i+1,': ',urls[i],sep='')
print()

mypool = Pool()  #multiple processes
start = mytime() #start the clock
#separate process for each url
results = mypool.map(myget, urls)
#print total elapsed
now = str(mytime() - start)
print("Total = " + now + " ms\n")
```

web11.py

We import from the `time` module so we can keep track of how long the code takes to run. We import from `urllib.request` to open and read web pages. Finally, we import from `multiprocessing` so we can do more than one thing at a time.

We define the same functions `mytime()` and `myget()` as in `web10.py`. We also use the same urls. We now print out the urls in order.

The next part of the code handles the parallelism. First, we create a `Pool` that keeps track of how many things we can do at once. This will vary depending on what kind of machine you run this on. The relevant bit of code is `mypool.map()`, which applies `myget()` to all the urls. Each of these applications of `myget()` proceeds in parallel, so if any page is slow to load, the other requests proceed unaffected. Since `myget()` prints out its run time as it finishes, you can see each process terminate. Finally, the program prints out the total run time.

If you run the code several times, you'll see that the total run time varies as a function of the speed of the internet, your connection, and the other machines you are interacting with.

You should see immediately that this program runs much more quickly than the `web10.py` program. You should also see that the run time for this code is *not* the sum of the length of the individual calls to `myget()`. How much faster it runs than that depends on your own computational resources, but the best-case scenario is that total run time is roughly equivalent to the slowest call to `myget()`.

Finally, you should also see that the order in which each call to `myget()` terminates is *not* necessarily the same order as in the initial list. Rather, assuming they all run in parallel, the order in which they finish is based on how long each one takes to complete. If your resources are limited, then you won't be able to sustain as many parallel processes, and this may not be quite true.

8.5 Unicode and Text Encoding

As linguists, we are often concerned with textual representations of language. The problem is that computers don't really process text directly. Rather, for computers, text is represented internally as numbers. This has consequences for us as programmers, and we deal with them in this section.

In the next sections, we discuss generally how characters are represented internally and the major character encodings you may run across. We then discuss methods for reading, writing, and converting different encodings. Finally, we discuss methods for deducing the encoding of some document or resource.

As above, the computer represents characters internally as numbers. In turn, numbers are represented as binary, or base-2, numbers.

Decimal	Binary
0	0
1	1
2	10
3	11
4	100
5	101
6	110
7	111
8	1000
9	1001
10	1010

Thus, we might imagine a system like the following for representing letters:

Decimal	Binary	Letters
0	0	a
1	1	b
2	10	c
3	11	d
4	100	e
5	101	f
6	110	g
7	111	h
8	1000	i
9	1001	j
10	1010	k
...	...	...

This is fine for individual letters, but presents a problem when we get to letter sequences. Using the hypothetical encoding above, a sequence like 101 is multiply ambiguous:

101	f
1+0+1	bab
10+1	cb

The problem is that the edges between encoded letters are not marked. One solution is to pad the numbers so they are of the same width, say, like this:

Decimal	Binary	Padded	Letters
0	0	0000	a
1	1	0001	b
2	10	0010	c
3	11	0011	d
4	100	0100	e
5	101	0101	f
6	110	0110	g
7	111	0111	h
8	1000	1000	i
9	1001	1001	j
10	1010	1010	k
...	...	...	

Doing this, the three interpretations for 101 above would be disambiguated like this:

ambiguous	letter sequence	unambiguous
101	f	0101
1+0+1	bab	000100000001
10+1	cb	00100001

The trick here is that we know every character has exactly four digits. This solution requires that we know how much padding needs to be added. In other words, we need to calculate from a finite set of characters what the "widest" binary number we will need is. From that, we can calculate how much padding to add. If we only needed to represent the letters above, we know that the biggest binary number we will need is 1010: four digits wide. Smaller numbers are padded accordingly. We refer to each of these digit slots as a *bit*.

When transmitting data encoded like this it is possible for errors to occur. For example, imagine you've got an encoding where every character is four digits wide, but the string of digits you're looking at is 43 digits wide. Something is wrong, but how do you find where the problem is? One strategy to detect errors is to suffix one more slot on each letter that helps detect errors: a *parity bit*.

One of the earliest approaches to character encoding is ASCII, and it has essentially this structure. It used 7 bits to encode each letter (plus an additional parity bit). An 8-bit unit of this sort is referred to as a *byte*.

If you do the math, you will see that with only seven binary bits to encode letters, this allows for a maximum of $2^7 = 128$ distinct characters. While this may suffice for a language like English, it doesn't do for all the characters of all the languages of the world. It certainly doesn't accommodate documents with multiple character sets.

Historically, at this point, we enter the wild west of character encoding. A huge variety of different approaches were adopted to deal with character sets beyond the 128 that ASCII could accommodate. Eventually, the *Unicode* standard was proposed, a uniform numerical value for every character in virtually every language. A growing majority of web resources that you will encounter make use of this standard.

There is an obvious issue, however. There are well over a million distinct characters in the Unicode standard. This means that if you want distinct bitwise representations of a constant width, each character will have to be substantially "wider" than 8 bits (1 byte).

There are two broad solutions to this that you will encounter. The first is *UTF-16* (Unicode Transformation Format), an encoding that bites the bullet and encodes (almost) every character with 16 bits (2 bytes).

The other more common solution is *UTF-8*. This approach encodes each character with a variable number of bytes: 1 to 4. The trick is that each byte is marked so that you can tell whether it is the start of some new multibyte character or a continuation of the multibyte sequence. The basic idea is as follows, where *x* indicates a bit that can be used for character values:

	Byte 1	Byte 2	Byte 3	Byte 4
1	0xxxxxxx			
2	110xxxxx	10xxxxxx		
3	1110xxxx	10xxxxxx	10xxxxxx	
4	11110xxx	10xxxxxx	10xxxxxx	10xxxxxx

The really interesting part of UTF-8, however, is that the one-byte case is identical to ASCII. That is, any ASCII-encoded text is a legal subcase of UTF-8. The most frequent encoding seen on the web today is UTF-8.

8.6 Bytes and Strings

Python is natively UTF-8. What this means is that file IO and the interactive environment handle UTF-8 characters directly. For example, we might enter the following in the interactive environment:

```
>>> x = 'русский язык'
>>> y = '中文'
>>> len(x)
12
>>> len(y)
2
```

Note that the Russian and Chinese characters are displayed properly and their lengths are calculated correctly. This also works from a program file:

```
x = 'русский язык'
y = '中文'
print(x,': ',len(x),sep='')
print(y,': ',len(y),sep='')
```

web12.py

Similarly, if a file is encoded as UTF-8, it can be read from directly. The following program exemplifies:

```
f = open('enc/russian.txt','r')
r = f.read()
f.close()
print('Some Russian:',r,end='')
```

```
f = open('enc/chinese.txt','r')
c = f.read()
f.close()
print('Some Chinese:',c,end='')
```

web13.py

In this case, the two files read from contain Russian and Chinese characters and are both encoded as UTF-8. (These files are included among the text program files.)

On the other hand, if a file is encoded in some other way, this will fail. Here we try to read Chinese from a file encoded in the *Big5* encoding, and it fails.

```
f = open('enc/cb5.txt','r')
f.read()
...
UnicodeDecodeError ...
```

A similar problem occurs if we try to read Russian from a file encoded in *Windows-1251* encoding:

```
f = open('enc/r1251.txt','r')
f.read()
...
UnicodeDecodeError ...
```

To read files in other encodings like this, we must first read in the raw bytes and then explicitly convert the encodings using the bytes method `decode()`. The following program exemplifies:

```
#open a file to read bytes
f = open('enc/cb5.txt','rb')
c = f.read()              #read the bytes
#convert from big5 to utf8 characters
c = c.decode('big5')
print('Chinese:',c)      #print characters

#open a file to read bytes
f = open('enc/r1251.txt','rb')
r = f.read()              #read the bytes
#convdert from windows-1251 to utf8
r = r.decode('1251')
print('Russian:',r)      #print characters
```

web14.py

In both cases, we invoke the open() function with the argument 'rb' so that we read raw bytes. The read() command is then executed without arguments. To display these strings we must then convert from bytes to UTF-8, and to do that we must explicitly convert from the relevant encodings with the decode() method.

If you want to write to a file in an encoding other than UTF-8, you must also use raw bytes. The second argument in this case is 'wb'. The following program writes out strings in UTF-8 converted to Windows-1251 and Big5 encodings using the bytes method encode().

```
r = 'русский язык'    #a string in Russian
#open a connection to write bytes
f = open('enc/rout1251.txt','wb')
#write bytes in windows-1251
f.write(r.encode('1251'))
f.close()              #close connection
#a string in chinese
c = '中文'
#open a connection to write bytes
f = open('enc/coutb5.txt','wb')
#write bytes in big5
f.write(c.encode('big5'))
f.close()              #close connection
```

web15.py

8.7 What Is the Encoding?

The biggest challenge to working with character encodings is that you may have a document you want to read in and simply not know what the encoding is. In this section, we discuss ways to deal with this.

The first "solution" to the encoding problem is to make sure that it's really a problem for you! In Section 8.8 we develop a webcrawler program. That program crawls the web looking for webpages in Welsh. It includes a function to do language identification that *assumes the page is encoded in UTF-8*. A side effect of this is that web pages in other encodings are assumed to be not in Welsh. This probably misses some pages that are in Welsh, but it also simplifies the language identification problem. Pages in other encodings are taken to be not in Welsh, *and that's generally true.*

The point is that – for that application – we don't need to figure out what encoding a page is in.

Another option for web pages specifically is that the encoding is specified in the document itself. In more recent versions of HTML this occurs in the `<meta>` tag. For example, the following in the head of an HTML document indicates that the document is encoded in UTF-8.

```
<head>
   <meta charset="UTF-8">
</head>
```

Unfortunately, this is not a required part of every HTML page. In addition, this doesn't help with determining the encoding of a document that is not in HTML.

Another solution is to attempt to figure out what the encoding is. The basic logic is that there are only a finite number of encodings out there, and they have specific structures in terms of what kinds of bytes and byte sequences occur. In addition, certain encodings are tailored to specific languages. Thus we can examine a document to see what encodings are possible and whether the text produced is consistent with the languages we expect to see.

One module that does this is `chardet` (not part of the default Python distribution). If we run this on our Russian file, it succeeds:

```
>>> import chardet
>>> f = open('enc/r1251.txt','rb')
>>> r = f.read()
>>> chardet.detect(r)
{'language': 'Russian',
  'encoding': 'windows-1251',
  'confidence': 0.99}
```

Here we import the `chardet` module and invoke its `detect()` function on bytes read from the short Russian file. That function returns a dictionary with three records indicating language, encoding, and its confidence in determining those.

This technique fails with our Chinese file:

```
>>> import chardet
>>> f = open('enc/cb5.txt','rb')
>>> c = f.read()
>>> chardet.detect(c)
{'language': '',
  'encoding': 'ISO-8859-1',
  'confidence': 0.73}
```

Presumably this is because the Chinese file is only two characters long and does not provide enough information for the `detect()` function. If we do this with longer downloaded web pages in the same languages and encodings, the function is successful in both cases:

```
>>> f = open('enc/rnd.html','rb')
>>> r = f.read()
>>> chardet.detect(r)
{'language': 'Russian',
  'encoding': 'windows-1251',
  'confidence': 0.9596062000298741}
```

```
>>> f = open('enc/dh.html','rb')
>>> c = f.read()
>>> chardet.detect(c)
{'language': 'Chinese',
  'encoding': 'Big5',
  'confidence': 0.99}
```

The `bs4` module also includes a method for guessing the encoding of a document. This method works in conjunction with `chardet` and filters away HTML markup to make it more effective. Using the same downloaded HTML pages as above, we get:

```
>>> from bs4 import UnicodeDammit
>>> dr = UnicodeDammit(r)
>>> dr.original_encoding
'windows-1251'
>>> dc = UnicodeDammit(c)
>>> dc.original_encoding
'big5'
```

8.8 A Webcrawler

In this section we program a simple webcrawler: a program that searches the web for webpages of interest and collects relevant data. In our case, we will search for webpages in Welsh and save the text.

As usual, we build our program up in stepwise fashion to model good programming habits. Let's start by reading in one page from a Welsh news site.

```
from urllib.request import urlopen

#welsh news site
url = 'http://golwg360.cymru'
w = urlopen(url,timeout=5)  #open connection
t = w.read()                #read whole page
#print number of words
print(len(t.split()))
```

web16.py

What we will ultimately want is a program that iterates through a list of urls. If a webpage satisfies some criterion – in this case, say, that it's in Welsh – then we save the text and add the links from that page to our list of links. We then continue. Let's add this general logic to our program.

```
from urllib.request import urlopen

#seed for list of urls
urls = ['http://golwg360.cymru',
'http://www.u.arizona.edu/~hammond']
res = []              #results
i = 1                 #iterate through list
while urls and i < 100:
   u = urls.pop(0)    #open/read url
   w = urlopen(u,timeout=5)
   t = w.read()
   #placeholder: count words
   print(i,': ',len(t.split()),sep='')
   i += 1
   #check if page is in Welsh
   if True:
      res.append([u,t])
      #extract links and append
```

web17.py

We converted our url to a list of two urls. I've deliberately put one clearly Welsh url in the list and one that is clearly not Welsh. We've also created a list to store the results of our crawling. We then iterate through the list as long as there are urls on the list and our counter doesn't reach 100. We will ultimately be adding links to the list and then removing them as we work through them. We need to check at each iteration that there are actually items still on the list. We add the counter so that the program stops at some point.

We then go through and read each page on our list. In this version of the program we print the length of the page, but that's just a placeholder. We will ultimately be checking each page to see if it's in Welsh. If it is, we save it and add its links to the list of urls.

Next, let's use `BeautifulSoup` to extract the text and the links.

```
from urllib.request import urlopen
from bs4 import BeautifulSoup
import re

#seed for list of urls
urls = ['http://golwg360.cymru',
'http://www.u.arizona.edu/~hammond']
res = []                    #results
i = 1                       #iterate through list
while urls and i < 100:
   u = urls.pop(0)          #open/read url
   w = urlopen(u,timeout=5)
   h = w.read()
   #parse html
   s = BeautifulSoup(h,'html.parser')
   t = s.body.get_text()
   i += 1
   #check if page is in Welsh
   if True:
      res.append([u,t])
      print(u,': ',len(t.split()),sep='')
      #extract links
      links = s.find_all('a')
      for l in links:
         print('\t',l.get('href'))
```

web18.py

We will ultimately want to follow links, so we need to be mindful of what kind of links we see here. There are two distinctions of interest. First, we have relative and absolute links. Absolute links begin with `http://` or `https://` and relative links do not. Relative links are, as you might expect, urls relative to the current page. For example, if we are on a page `http://www.where.edu` and we find the relative link `/wales.html`, we would need to convert it to `http://www.where.edu/wales.html`. Thus, if we want to follow relative links, we will need to convert them to absolute links.

Another kind of link is a local link, which can occur in either an absolute or a relative link. These include the # symbol and might look like `wales.html#yma` or `http://www.where.edu#lle`. These link to the same pages as the two preceding links, but position the browser window in specific locations. If we are interested only in the content of web pages, we should prune the local link information.

The next version of the code strips out local links and converts relative links to absolute links.

```
from urllib.request import urlopen
from bs4 import BeautifulSoup
import re

#fix relative and local links
def fixlinks(u,l):
   res = re.sub('#.*$','',l)
   m = re.search('^http',res)
   if m:
      return res
   res = u + '/' + res
   res = re.sub('([^:]/)/+','\\1',res)
   return res

#seed for list of urls
urls = ['http://golwg360.cymru',
'http://www.u.arizona.edu/~hammond']
res = []              #results
i = 1                 #iterate through list
while urls and i < 100:
   u = urls.pop(0)  #open/read url
   w = urlopen(u,timeout=5)
   h = w.read()
   #parse html
   s = BeautifulSoup(h,'html.parser')
   t = s.body.get_text()
   i += 1
   #check if page is in Welsh
   if True:
      res.append([u,t])
```

```
        print(u,': ',len(t.split()),sep='')
        #extract links
        links = s.find_all('a')
        for l in links:
            link = l.get('href')
            if link:
                print('\t',link)
                fixedlink = fixlinks(u,link)
                print('\t\t',fixedlink)
```

web19.py

The next issue to sort out is what happens if a link doesn't work or times out. The code as currently written stops if that occurs. What would be better is if the code simply skips over bad links. We address this with a new control structure `try/except`. Here commands in the `try` block are executed. If no error occurs, the program continues. If an error does occur, the commands in the `except` block apply. Here's a simple example:

```
print('Before the first try block:')
try:
    print('\tthree' + 3)
except:
    print("\tThat math doesn't work.")
print('Before the second try block:')
try:
    print('\tthree' + ' + 3.')
except:
    print("That math doesn't work.")
print('All done.')
```

web20.py

Here the first `try` block fails because + is not defined for strings and numbers together and its `except` block applies. The second `try` block does succeed, so its `except` block does not apply.

Let's now use this in our webcrawler code:

```
from urllib.request import urlopen
from bs4 import BeautifulSoup
import re
```

```
#fix relative and local links
def fixlinks(u,l):
   res = re.sub('#.*$','',l)
   m = re.search('^http',res)
   if m:
      return res
   res = u + '/' + res
   res = re.sub('([^:]/)/+','\\1',res)
   return res

#seed for list of urls
urls = ['http://golwg360.cymru',
'http://bad',
'http://www.u.arizona.edu/~hammond']
res = []                #results
i = 1                   #iterate through list
while urls and i < 100:
   u = urls.pop(0)      #open/read url
   i += 1
   try:
      w = urlopen(u,timeout=5)
      h = w.read()
   except:
      print('bad url:',u)
      continue
   #parse html
   s = BeautifulSoup(h,'html.parser')
   t = s.body.get_text()
   #check if the page is in Welsh
   if True:
      res.append([u,t])
      print(u,': ',len(t.split()),sep='')
      #extract links
      links = s.find_all('a')
      for l in links:
         link = l.get('href')
         if link:
            print('\t',link)
            fixedlink = fixlinks(u,link)
            print('\t\t',fixedlink)
```

web21.py

Here if the url fails to open or we can't read from it, we print out the name of the url and go on to the next one.

Another issue we have to face is that we may encounter nontextual data. Sometimes it will be apparent from the url, i.e., it ends in `.au`, `.zip`, or some other suffix. Sometimes, however, it will not. To accommodate this, we use the same conversion string conversion utility as above: `decode('UTF-8')`. This will succeed if the webpage is appropriate text. The following version of the code adds this.

```
from urllib.request import urlopen
from bs4 import BeautifulSoup
import re

#fix relative and local links
def fixlinks(u,l):
   res = re.sub('#.*$','',l)
   m = re.search('^http',res)
   if m:
      return res
   res = u + '/' + res
   res = re.sub('([^:]/)/+','\\1',res)
   return res

#seed for list of urls
urls = ['http://golwg360.cymru',
'http://bad',
'http://www.u.arizona.edu/~hammond/greeting.au',
'http://www.u.arizona.edu/~hammond']
res = []                    #results
i = 1                       #iterate through list
while urls and i < 100:
   u = urls.pop(0)          #open/read url
   i += 1
   try:
      w = urlopen(u,timeout=5)
      h = w.read()
      h = h.decode('UTF-8')
   except:
      print('bad url:',u)
      continue
   #parse html
   s = BeautifulSoup(h,'html.parser')
```

```
    t = s.body.get_text()
    #check if the page is in Welsh
    if True:
        res.append([u,t])
        print(u,': ',len(t.split()),sep='')
print('Stored pages:',len(res))
```

web22.py

Our next step is to implement the check for whether the page is in Welsh. There are sophisticated ways to do this, but we will use an intuitive simple strategy. First, we calculate what are the most common words of Welsh. We then score documents by what percentage of the document are potentially instances of these words.

First, we use the CEG corpus[4] to find the 20 most common word forms of Welsh. We set aside forms marked with apostrophes and forms with accents.

yn	y	i	a	o
ei	ar	yr	bod	ac
am	wedi	hi	ond	eu
fel	na	un	ni	mewn

We now write a function that computes what percentage of a webpage's words are on this list. Our goal is to find a separation point between pages in Welsh and pages in other languages.

```
from urllib.request import urlopen
from bs4 import BeautifulSoup
import re

welshwords = ['yn','y','i','a','o',
'ei','ar','yr','bod','ac',
'am','wedi','hi','ond','eu',
'fel','na','un','ni','mewn']

def welsh(u,t):
    n = t.lower()
    n = re.sub('[^a-z]',' ',n)
    n = re.sub('  +',' ',n)
    wds = n.split()
    total = len(wds)
```

[4] Ellis, N. C., C. O'Dochartaigh, W. Hicks, M. Morgan, and N. Laporte. 2001. *Cronfa Electroneg o Gymraeg (CEG)*, www.bangor.ac.uk/canolfanbedwyr/ceg.php.en

```
   wcount = 0
   for w in wds:
      if w in welshwords:
         wcount += 1
   percent = wcount/total
   print(u,'\n\t',wcount,'/',total, \
      ' (',percent,')',sep='')
   return True

#fix relative and local links
def fixlinks(u,l):
   res = re.sub('#.*$','',l)
   m = re.search('^http',res)
   if m:
      return res
   res = u + '/' + res
   res = re.sub('([^:]/)/+','\\1',res)
   return res

#seed for list of urls
urls = ['http://golwg360.cymru',
'https://cy.wikisource.org/wiki/Hafan',
'http://haciaith.com',
'http://techiaith.cymru',
'https://www.bbc.co.uk/cymru',
'https://www.yahoo.com',
'https://news.google.com/news/',
'https://en.wikipedia.org/wiki/Main_Page',
'http://www.u.arizona.edu/~hammond']
res = []                  #results
i = 1                     #iterate through list
while urls and i < 100:
   u = urls.pop(0)        #open/read url
   i += 1
   try:
      w = urlopen(u,timeout=5)
      h = w.read()
      h = h.decode('UTF-8')
   except:
      print('bad url:',u)
      continue
   #parse html
```

```
    s = BeautifulSoup(h,'html.parser')
    t = s.get_text()
    #check if the page is in Welsh
    if welsh(u,t):
        res.append([u,t])
        print('\t',len(t.split()),sep='')
print('Stored pages:',len(res))
```

web23.py

We've changed our list of urls here so that we have pages we know are in Welsh and others that we know are in English. What we see when we run this is that when the words above comprise 9 percent or more of the words on the page, we can be pretty sure the page is in Welsh. This misses some pages, but it's a reasonable starting point. The following code alters the `welsh()` function to do this.

```
from urllib.request import urlopen
from bs4 import BeautifulSoup
import re

#most frequent Welsh words
welshwords = ['yn','y','i','a','o',
'ei','ar','yr','bod','ac',
'am','wedi','hi','ond','eu',
'fel','na','un','ni','mewn']

#function to test for Welsh
def welsh(u,t):
    n = t.lower()
    n = re.sub('[^a-z]',' ',n)
    n = re.sub(' +',' ',n)
    wds = n.split()
    total = len(wds)
    wcount = 0
    for w in wds:
        if w in welshwords:
            wcount += 1
    percent = wcount/total
    if percent > .08:
        return True
    else:
        return False
```

```
#fix relative and local links
def fixlinks(u,l):
   res = re.sub('#.*$','',l)
   m = re.search('^http',res)
   if m:
      return res
   res = u + '/' + res
   res = re.sub('([^:]/)/+','\\1',res)
   return res

#seed for list of urls
urls = ['http://golwg360.cymru',
'https://cy.wikisource.org/wiki/Hafan',
'http://haciaith.com',
'http://techiaith.cymru',
'https://www.bbc.co.uk/cymru',
'https://www.yahoo.com',
'https://news.google.com/news/',
'https://en.wikipedia.org/wiki/Main_Page',
'http://www.u.arizona.edu/~hammond']
res = []                  #results
i = 1                     #iterate through list
while urls and i < 100:
   u = urls.pop(0)        #open/read url
   i += 1
   try:
      w = urlopen(u,timeout=5)
      h = w.read()
      h = h.decode('UTF-8')
   except:
      print('bad url:',u)
      continue
   #parse html
   s = BeautifulSoup(h,'html.parser')
   t = s.get_text()
   #check if page is in Welsh
   if welsh(u,t):
      res.append([u,t])
      print(u,': ',len(t.split()),sep='')
print('Stored pages:',len(res))
```

web24.py

The next step is to add urls to our list if the page they are on is in Welsh. The following code does this:

```
from urllib.request import urlopen
from bs4 import BeautifulSoup
import re

#most frequent Welsh words
welshwords = ['yn','y','i','a','o',
'ei','ar','yr','bod','ac',
'am','wedi','hi','ond','eu',
'fel','na','un','ni','mewn']

#function to test for Welsh
def welsh(u,t):
   n = t.lower()
   n = re.sub('[^a-z]',' ',n)
   n = re.sub('  +',' ',n)
   wds = n.split()
   total = len(wds)
   wcount = 0
   for w in wds:
      if w in welshwords:
         wcount += 1
   percent = wcount/total
   if percent > .08:
      return True
   else:
      return False

#fix relative and local links
def fixlinks(u,l):
   res = re.sub('#.*$','',l)
   m = re.search('^http',res)
   if m:
      return res
   res = u + '/' + res
   res = re.sub('([^:]/)/+','\\1',res)
   return res

#seed for list of urls
urls = ['http://golwg360.cymru',
```

```
'https://cy.wikisource.org/wiki/Hafan',
'http://haciaith.com',
'http://techiaith.cymru',
'https://www.bbc.co.uk/cymru',
'https://www.yahoo.com',
'https://news.google.com/news/',
'https://en.wikipedia.org/wiki/Main_Page',
'http://www.u.arizona.edu/~hammond']
res = []                    #results
i = 1                       #iterate through list
while urls and i < 20:
   u = urls.pop(0)          #open/read url
   i += 1
   try:
      w = urlopen(u,timeout=5)
      h = w.read()
      h = h.decode('UTF-8')
   except:
      print('bad url:',u)
      continue
   #parse html
   s = BeautifulSoup(h,'html.parser')
   t = s.get_text()
   #check if the page is in Welsh
   if welsh(u,t):
      res.append([u,t])
      print(u,': ',len(t.split()),sep='')
      links = s.find_all('a')
      for l in links:
         lu = l.get('href')
         if lu:
            urls.append(fixlinks(u,lu))
print('Stored pages:',len(res))
```

web25.py

Notice that we must test if we were actually able to extract the `href` attribute from the `a` tag before trying to append it to our list.

This works fine but fails in an important respect. The program so far adds links regardless of whether they are already in the list or we have already looked at them and popped them off the list. The following code implements these checks.

```
from urllib.request import urlopen
from bs4 import BeautifulSoup
import re

#most frequent Welsh words
welshwords = ['yn','y','i','a','o',
'ei','ar','yr','bod','ac',
'am','wedi','hi','ond','eu',
'fel','na','un','ni','mewn']

#function to test for Welsh
def welsh(u,t):
   n = t.lower()
   n = re.sub('[^a-z]',' ',n)
   n = re.sub('  +',' ',n)
   wds = n.split()
   total = len(wds)
   wcount = 0
   for w in wds:
      if w in welshwords:
         wcount += 1
   percent = wcount/total
   if percent > .08:
      return True
   else:
      return False

#fix relative and local links
def fixlinks(u,l):
   res = re.sub('#.*$','',l)
   m = re.search('^http',res)
   if m:
      return res
   res = u + '/' + res
   res = re.sub('([^:]/)/+','\\1',res)
   return res

#seed for list of urls
urls = ['http://golwg360.cymru',
'https://cy.wikisource.org/wiki/Hafan',
'http://haciaith.com',
'http://techiaith.cymru',
```

```
    'https://www.bbc.co.uk/cymru',
    'https://www.yahoo.com',
    'https://news.google.com/news/',
    'https://en.wikipedia.org/wiki/Main_Page',
    'http://www.u.arizona.edu/~hammond']
res = []                #results
already = []            #urls already checked
i = 1                   #iterate through list
while urls and i < 20:
    u = urls.pop(0)     #open/read url
    already.append(u)
    i += 1
    try:
        w = urlopen(u,timeout=5)
        h = w.read()
        h = h.decode('UTF-8')
    except:
        print('bad url:',u)
        continue
    #parse html
    s = BeautifulSoup(h,'html.parser')
    t = s.get_text()
    #check if the page is in Welsh
    if welsh(u,t):
        res.append([u,t])
        print(u,': ',len(t.split()),sep='')
        links = s.find_all('a')
        for l in links:
            lu = l.get('href')
            if lu:
                lufixed = fixlinks(u,lu)
                if lu not in already \
                    and lu not in urls:
                    urls.append(lufixed)
print('Stored pages:',len(res))
```

web26.py

Finally, the program must save results when it's done. The following code adds this feature:

```
from urllib.request import urlopen
from bs4 import BeautifulSoup
```

```
import re

#most frequent Welsh words
welshwords = ['yn','y','i','a','o',
'ei','ar','yr','bod','ac',
'am','wedi','hi','ond','eu',
'fel','na','un','ni','mewn']

#function to test for Welsh
def welsh(u,t):
   n = t.lower()
   n = re.sub('[^a-z]',' ',n)
   n = re.sub('  +',' ',n)
   wds = n.split()
   total = len(wds)
   wcount = 0
   for w in wds:
      if w in welshwords:
         wcount += 1
   percent = wcount/total
   if percent > .08:
      return True
   else:
      return False

#fix relative and local links
def fixlinks(u,l):
   res = re.sub('#.*$','',l)
   m = re.search('^http',res)
   if m:
      return res
   res = u + '/' + res
   res = re.sub('([^:]/)/+','\\1',res)
   return res

#seed for list of urls
urls = ['http://golwg360.cymru',
'https://cy.wikisource.org/wiki/Hafan',
'http://haciaith.com',
'http://techiaith.cymru',
'https://www.bbc.co.uk/cymru',
'https://www.yahoo.com',
```

```
'https://news.google.com/news/',
'https://en.wikipedia.org/wiki/Main_Page',
'http://www.u.arizona.edu/~hammond']
res = []                 #results
already = []             #urls already checked
i = 1                    #iterate through list
while urls and i < 20:
   u = urls.pop(0)       #open/read url
   already.append(u)
   i += 1
   try:
      w = urlopen(u,timeout=5)
      h = w.read()
      h = h.decode('UTF-8')
   except:
      print('bad url:',u)
      continue
   #parse html
   s = BeautifulSoup(h,'html.parser')
   t = s.get_text()
   #check if the page is in Welsh
   if welsh(u,t):
      res.append([u,t])
      print(u,': ',len(t.split()),sep='')
      links = s.find_all('a')
      for l in links:
         lu = l.get('href')
         if lu:
            lufixed = fixlinks(u,lu)
            if lu not in already \
               and lu not in urls:
                  urls.append(lufixed)
print('Stored pages:',len(res))
#save results
f = open('already.txt','w')  #already
for u in already:
   f.write(u+'\n')
f.close()
f = open('urls.txt','w')     #urls
for u in urls:
   f.write(u+'\n')
f.close()
```

```
f = open('resuts.txt','w')    #results
f.write('<results>\n')
for r in res:
   f.write('<record>\n')
   u = r[0]
   t = r[1]
   f.write('<url>\n')
   f.write(u+'\n')
   f.write('</url>\n')
   f.write('<text>\n')
   f.write(t+'\n')
   f.write('</text>\n')
   f.write('</record>\n')
f.write('</results>\n')
f.close()
```

web27.py

The code simply writes the three data structures to files: `urls` to `urls.txt`, `already` to `already.txt`, and `res` to `results.txt`. The only complication is that the `res` list is a list of lists that we want to make easy use of.

To allow for this we've written the `results.txt` file as an *XML* file. XML is a language-independent general format for structuring data of all sorts. The basic idea is that we have tags, much like in HTML, but the tags can be whatever we want. In our case, we have tags for `<results>`: the whole file, `<record>`: a url-text pair, `<url>`: the url, and `<text>`: the text of the url. Here is a schematic view:

```
<results>
  <record>
   <url>
   ...
   </url>
   <text>
   ...
   </text>
  </record>
  ...
</results>
```

In fact, the bs4 module has simple functions for dealing with XML data, which we leave as an exercise. There are, in fact, a number of other things that can be done to improve this program, but all of them are left as exercises.

8.9 Exercises

8.1 Augment the webcrawler code so that you can restart the system with existing files. The basic idea is that you might want to add urls and additional data.

8.2 The webcrawler is slow. Rewrite it so that it uses parallel techniques and runs faster. This is hard!

8.3 The webcrawler is designed to search for Welsh webpages. Convert it to a different language.

8.4 The `welsh()` function in the webcrawler program can be revised in many ways. Can you come up with another approach? Design a test program so that your new function can be compared against the one in the text.

8.5 Write a function that takes a url as an argument and returns any items on the page marked with the `<strong>` tag.

8.6 Write a function that takes a url as an argument and extracts and prints any ordered lists marked with `<ol>`. Note that individual items in such a list are marked with `<li>`. Also, make sure each item in the list is clearly separated *and numbered appropriately*. Use *BeautifulSoup* to do this.

8.7 Do the same thing as in the previous exercise but without using `BeautifulSoup`.

8.8 Write a function that takes a webpage in some encoding as an argument. Your function should apply the `chardet.detect()` function to the page *incrementally* and report back how much of the page in bytes had to be read to reach a certain confidence level (that you specify) to identify the encoding.

8.9 Do the same thing as in the previous exercise but use `UnicodeDammit`. Which does better?

8.10 Choose a language and do some research on what encodings are typically used for it on the web. Then find examples of each on the web.

8.11 Tweak the webcrawler program from the preceding chapter so that it reports whether it finds the `<meta charset=...>` tag and what value it returns. Your code should save this as a separate field in the XML results file.

8.12 What special challenges occur with respect to text encoding when a language is written right to left? Explain the problem and show how it can be solved with Python.

8.13 **Web**: Snoop around the web to find examples of at least one character encoding that we have not exemplified here. Find a web page that uses the encoding and write a program that loads the page, handles the encoding properly, and builds and displays a concordance.

9 Objects

In this chapter we introduce the general logic and structure of *object-oriented (OO) programming*. So far, we have treated programs as sequences or groups of commands. In this chapter, we will instead treat programs as networks of objects or things.

Some tasks really lend themselves to this sort of approach, others less so. Regardless, OO programming forces programmers to think about their programs differently and allows for separating the different parts of a program in a different way as well.

We first introduce the general logic of classes and objects. We then treat the basic syntax of classes, focusing on the difference between classes and instances of classes. We then treat the mechanism of inheritance. Finally, we conclude with an extended example: syllabification.

9.1 General Logic

How do we go about treating a programming problem as a network of objects rather than a sequence of commands? Let's take a concrete example. Imagine we want to read in some text file and parse it into sentences. Up to this point, we might do it like this:

```
import re

#open and read file
f = open('alice.txt','r')
t = f.read()
f.close()
t = t[10841:]        #strip header
#split into sentences
ss = re.split('([\.?!])',t)
i = 0                #show first 10
while i < 20:
   s = ss[i] + ss[i+1] + '\n'
```

```
    print(s)
    i += 2
```

obj1.py

Here we take advantage of the fact that when we use `re.split()` and mark the match with parentheses, the splitting string is returned as well. This returns a list of the sentences alternating with the terminating punctuation marks.

We might instead conceptualize this differently. We want to take a `File` object, which represents the file; create a `Text` object, which represents the text, from that; and then extract `Sentence` objects from it. At the highest level, this might look like this:

```
#Only schematic; generates errors!

f = File('alice.txt')

t = Text(f)

ss = t.getSentences()

for s in ss[:10]:
   print(s)
```

obj2.py

Ultimately, the same work has to be done in both cases. The difference is in how we organize the code.

In general, we'll see that OO code, on the one hand, tends to be longer. On the other hand, it can often have a clearer logic. This means that when choosing whether to program in OO style, you often are choosing between writing less code that is less clear or more code that is more clear.

We come back to this program below after we've introduced the logic and syntactic sugar for objects.

9.2 Classes and Instances

When we create objects, we must distinguish the general from the specific. Thus, if we were to create word objects, we would want a general characterization of what a word is and then specific instances of that. The terminology used is that we create a general *class* of which we can have *instances*. At the simplest level, we might define a word class like this:

```
class Word:      #a class definition
   #this class has a variable
   info = "I'm a word."

w = Word()       #create instance of class
#access variable from instance
print('w:',w.info)
#access variable from class
print('Word:',Word.info)
```

obj3.py

We define a class with the keyword `class` followed by the class name and a colon. Class names are customarily capitalized. Properties of that class occur in the following indented block. These properties include variables and methods. In our `Word` class, we have only the variable `info`. A class is instantiated by invoking its name with parentheses. We can then extract the variable in the class body by suffixing that variable name to the instance `w` or to the class name `Word` itself. A variable like this that can be accessed from the class name or from any specific instance is referred to as a *class variable.*

Class definitions can include function definitions. A function defined in a class definition is a *method.* Here's a simple example:

```
class Word:                       #class def
   def whatami():                 #method def
      return "I'm a word."        #method body

#invoke method from class name
print('Word:',Word.whatami())
```

obj4.py

The method definition uses the keyword `def` just like a function definition, except that a method definition is *within* a class definition.

We invoke the method by suffixing it to the class name with a period and adding parentheses. This does *not* work with an instance of the `Word` class. Thus an expression like `w.whatami()` would generate an error. The problem is that `w.whatami()` is actually equivalent to something we might think of as `w.whatami(w)`. When you invoke a method from an instance of the class, it's as if the first argument to the method is the instance itself. Here's a version of the same program that uses this.

```
class Word:                  #class definition
   def whatami(self):        #instance method
      return "I'm a word."

#create instance of class
w = Word()
#call instance method
print('w:',w.whatami())
```

obj5.py

Here we define the `whatami()` method to take an argument that is never used. Now, when we call that method on the `w` instance of the class `Word`, everything works fine. It is customary to use the variable name `self` in just this case to indicate a *variable that is bound to the instance of the class*. In this case then, `self` is effectively a reference to `w`. In this case, we say that `whatami()` is an *instance method* because it refers to the particular instantiation of `Word` indirectly through the variable `self`.

We have seen class methods, instance methods, and class variables. There are *instance variables* too. These are created and accessed in a fashion similar to that for instance methods: by invoking the particular instance. The trick is that instance variables are created and accessed only through instance methods. Here's an example:

```
class Word:  #class definition
   #instance method
   def whatami(self):
      #creating an instance variable
      self.x = "I'm a noun!"

w = Word()    #create instance of class
w.whatami()   #call the instance method
print(w.x)    #print the instance variable
```

obj6.py

Here we define the class `Word` to include the instance method `whatami()`. We know it's an instance method because it is defined with the customary instance variable `self`. When that method is invoked, it creates an instance variable `self.x` and assigns it a string value. We know that that's an instance variable because it is defined in an instance method and has the prefix `self`. We instantiate our class in the usual way and call the instance method `whatami()`. Since it is an instance method, it must be invoked with the instantiated variable as a prefix. Finally, we print out the value of the instance variable, again using the instantiated variable as a prefix.

This is a fair bit of terminology, so let's review and make sure we understand the terms and concepts.

Class definition. A class definition specifies a type of object. When specified, it has the word `class` followed by a (capitalized) word, a colon, and then an indented block.

```
class MyClass:
    ...
```

Object or instance of a class. An object is an instance of some class. It is created by invoking the class name followed by parentheses.

```
x = MyClass()
```

Class method. A class method is a method specific to some class, but not specific to any instance of that class. It is defined in the body of a class definition and it does *not* refer to `self`. It is invoked with the class names as a prefix. For example:

```
class MyClass:
    def myMethod():
        print('wow')

MyClass.myMethod()
```

Instance method. An instance method is specific to some instance of a class. It is defined in a class definition body and includes `self` as its first argument. It is invoked with an instance of the class as a prefix. For example:

```
class MyClass:
    def myMethod(self):
        print('wow')

x = MyClass()
x.myMethod()
```

Class variable. A class variable is information relevant for an entire class. It is defined in the body of the class definition. It is invoked with the class name as a prefix. For example:

```
class MyClass:
   myVariable = 'wow'

print(MyClass.myVariable)
```

Instance variable. Finally, an instance variable is information relevant for some particular instance of a class. It is always created and manipulated through instance methods. It will always have `self` as a prefix in an instance method. If it is accessed directly from outside the class, it will have a class instance as a prefix. For example:

```
class MyClass:
   def setVal(self,x):
      self.n = x
   def getVal(self):
      return self.n

x = MyClass()
x.setVal(3)
print(x.getVal())
```

obj7.py

This last example shows an additional, not unexpected feature we haven't seen yet: instance methods may take other arguments as well. Here we define two instance methods, `setVal()` and `getVal()`. The `setVal()` method takes two arguments, `self` and one other, and creates an instance variable `n` with the value of its second argument. The other method `getVal()` retrieves the value of that instance variable.

Let's now make sense of the distinction between instance and class variables. Imagine we want to extend our `Word` class to distinguish parts of speech and to keep track of how many words we have. We might do that as follows:

```
class Word:                 #class definition
   count = 0                #class variable
   #instance method
   def setPOS(self,x):
      #create instance variable and
      #set to x
      self.pos = x
      #increment class variable count
      Word.count += 1
```

```
a = Word()              #make class instance
#set its instance variable to 'noun'
a.setPOS('noun')
b = Word()              #make another instance
#set its instance variable to 'verb'
b.setPOS('verb')
#print the value of count associated with a
print('a count:',a.count)
#print the value of count associated with b
print('b count:',b.count)
#print value of count associated with the class
print('Word count:',Word.count)
print('a POS:',a.pos) #print pos of a
print('b POS:',b.pos) #print pos of b
```

obj8.py

Here we define a class `Word` with a class variable `count` and an instance method `setPOS()`. The latter creates an instance variable `pos` and increments the value of `count`. We then instantiate the class twice for a noun and a verb. We then print the value of `count` associated with each instance and the class and the value of `pos` associated with each instance. As we expect, the values for `pos` are different, but the values for `count` are not; it is 2 in both cases.

Here, the class variable `count` keeps track of information relevant to the entire class: how many instances of it are there. On the other hand, the instance variable `pos` keeps track of information relevant to a specific instance of a class: its own part of speech. This is a fairly typical structure and logic.

There is a more customary way to implement something like the above. You can define your class so that when you instantiate it, it does some sort of automatic setting up. When a class is instantiated, Python automatically invokes a specific instance method if it exists: `__init__()`. If you include a definition for that method in your class definition, then that function will be run when the class is instantiated. Since it is an instance method, its first or only argument is `self`. Here's a simple example:

```
class MyClass:
   def __init__(self):
      print('Instantiating MyClass!')

x = MyClass()
y = MyClass()
```

obj9.py

We can do more interesting work with this, however. The following is a revision of `obj8.py`:

```
class Word:              #class definition
   count = 0             #class variable
   #initializer takes an extra arg
   def __init__(self,x):
      #create instance variable, set to x
      self.pos = x
      #increment class count variable
      Word.count += 1

a = Word('noun')     #instantiate with 'noun'
b = Word('verb')     #instantiate with 'verb'
#continue as before
print('a count:',a.count)
print('b count:',b.count)
print('a POS:',a.pos)
print('b POS:',b.pos)
```

obj10.py

Here we've defined the `__init__()` method to take an extra argument, which we use to set a value for the instance variable `pos`. We also increment the class variable `count`. Now we instantiate the class twice with different arguments. We get the same results when we print the contents of the variables associated with the two objects.

Let's now return to our example `obj2.py` on page 207. We can now flesh that out. We start by setting up preliminary class definitions so that the code at least runs:

```
class File:              #class def for File
   #initialize with arg
   def __init__(self,s):
      #instance variable for filename from arg
      self.filename = s
   #method that will eventually extract text
   def getText(self):
      return 'some text'

class Text:              #class def for Text
   #initialize by getting text from File arg
```

```
    def __init__(self,f):
        self.strings = f.getText()
    #method that will eventually
    #return a list of sentences
    def getSentences(self):
        return ['1','2','3','4','5']

#instantiate File with filename
f = File('alice.txt')
#instantiate Text with file instance
t = Text(f)
#extract sentences from Text
ss = t.getSentences()
for s in ss[:3]:     #print the first 3
    print(s)
```

obj11.py

We start with a `File` class. We know that it has to be created with a string argument that represents the name of the file to be read. We therefore set up an `__init__()` method that takes a string argument and saves it as an instance variable `filename`.

The `File` object will be an argument to the `Text` object, so we create a function `getText()` that `Text` will call to extract the text from the `File` object.

We next create a `Text` object. It has to be instantiated with the `File` object as an argument, so we define an `__init__()` method that takes a `File` object as an argument. It then invokes that `File` object's `getText()` method to extract the text of the file and assigns that to an instance variable `strings`.

Finally, we know that we want to extract sentences from the `Text` object with a method `getSentences()`, so we create an instance method with that name that returns a list of dummy sentences.

This code runs, but does nothing interesting yet. It is the frame that we will build our code on. Note that the code that invokes these classes – in the last few lines of the program – is fairly simple and that the real work of reading files, extracting text, etc., will be done "behind the scenes" in the definitions of the classes.

Let's now flesh out the `File` class to actually open the file, read in the text, and remove the header.

```
class File:              #File class def
    #initialize with string argument
    def __init__(self,s):
```

```
        #open connection to file
        f = open(s,'r')
        #read it all in and assign to
        #instance variable
        self.text = f.read()
        f.close()    #close connection
        #strip header
        self.text = self.text[10841:]
        #temp placeholder message!
        print('<file read in>')
    #instance method to return the text
    def getText(self):
        return self.text

class Text:          #Text class def
    #initialize with File arg
    def __init__(self,f):
        #set instance variable
        self.strings = f.getText()
    #return sentences
    def getSentences(self):
        return ['1','2','3','4','5']

#create an instance of File
f = File('alice.txt')
#create Text instance from File instance
t = Text(f)
#get the sentences
ss = t.getSentences()
for s in ss[:3]:   #print the first 3
    print(s)
```

obj12.py

Here we've fleshed out the `__init__()` method for `File` to read the file and strip the header. We've also added some temporary code to print out that this initialization has happened. We will of course delete this when we're done.

Let's now flesh out `Text`:

```
import re

class File:          #File def
```

```
   #initialize with file name
   def __init__(self,s):
      #read in file
      f = open(s,'r')
      self.text = f.read()
      f.close()
      #strip header
      self.text = self.text[10841:]
   #return text of the file
   def getText(self):
      return self.text

class Text:          #Text class def
   #initialize with File instance arg
   def __init__(self,f):
      #get the text from File instance
      text = f.getText()
      #split on sentence breaks
      #saving split letter with parens!
      bits = re.split('([\.!?])',text)
      #instance variable list to save sentences
      self.sentences = []
      #assemble each sentence and add to list
      i = 0
      while i < len(bits)-1:
         self.sentences.append(
            bits[i]+bits[i+1]
         )
         i += 2
   #return the sentences
   def getSentences(self):
      return self.sentences

#instance of File
f = File('alice.txt')
t = Text(f)          #instance of Text
#get sentences
ss = t.getSentences()
for s in ss[:10]:  #print the first 10
   print(s)
```

obj13.py

We first transfer the text from the `File` object via its `getText()` method to a variable `text` in the `Text` object. This variable has no prefix, so it is entirely local to this function. We then split the text into sentence-sized units with `re.split()`. Since we've marked the splitting elements with parentheses, they are also in the resulting list, so we then concatenate sentences and their final punctuation in pairs and put them in an instance variable `sentences`. Since this is prefixed with `self`, it is preserved in the instance outside the `__init__()` function and available to the `getSentences()` instance method.

9.3 Inheritance

When you start writing fairly large programs in OO style, you end up with classes that are related. Python allows you to express and simplify these relationships with *inheritance*. The basic idea is that if some class is specified to inherit from some other class, then the methods and variables of that latter class are available to the former. The syntax for this is for the inheriting class to specify the classes it inherits from in parentheses after the class name in the class definition. For example:

```
class MyParent:
   ...

class MyChild(MyParent):
   ...
```

Here the methods and variables of `MyParent` are available from `MyChild`. Here's a working example:

```
class MyParent:       #parent class def
   def wow():         #class method
      print('wow!')
   #instance method with arg
   def printthis(self,x):
      print(x)

#child class def inherits from MyParent
class MyChild(MyParent):
   pass               #nothing in the body

a = MyChild()         #make instance of child
#inherits parent class method
```

```
MyChild.wow()
#inherits parent instance method
a.printthis('oh?')
```

obj14.py

Here we define a class MyParent with a class method wow() and an instance method printthis(). We then define a class MyChild that inherits from MyParent and has no methods or variables itself; we indicate an empty class body with pass. We can call the methods of MyParent from MyChild as if they were part of MyChild.

The same is true of variables:

```
class MyParent:          #parent class def
   info = 'wow!'         #class variable
   def oh(self):         #instance method
      #creates an instance variable
      self.oh = 'yippee!'

#child class def
class MyChild(MyParent):
   pass                  #nothing in the body

#create instance of child
a = MyChild()
#inherits parent class variable
print(MyChild.info)
#inherits parent instance method
a.oh()
#inherits parent instance variable
print(a.oh)
```

obj15.py

Here we've defined a class variable info for MyParent. We've also created an instance variable for it via the instance method oh(). These are all available to the inheriting class MyChild. First we instantiate that class as a. We then print out the value of info, inherited from MyParent. We then run the oh() instance method, also inherited from MyParent. That creates an instance variable oh in a, which we access in the usual way.

Inheritance is transitive as well. The methods and variables of classes more than one "generation" back are also available. For example:

```
class MyGrandparent: #grandparent class def
   info = 'wow'      #class variable
   def hm(self,x):   #instance method
      self.oh = x    #instance variable

#parent class inherits from MyGrandparent
class MyParent(MyGrandparent):
   pass              #nothing in body

#child class inherits from MyParent
class MyChild(MyParent):
   pass              #nothing in body

#an instance of the child class
a = MyChild()
#inherits grandparent class variable
print(a.info)
#inherits grandparent instance method
a.hm('bummer')
#inherits grandparent instance variable
print(a.oh)
```

obj16.py

Here we define a class `MyGrandparent`. The class `MyParent` inherits from that, and the class `MyChild` inherits from `MyParent`. Methods and variables of `MyGrandparent` are then available to `MyChild`.

You can override inherited methods and variables by defining them locally. Here's a simple example:

```
class MyParent:       #parent class def
   info = 'wow'       #class variable
   #instance method
   def mySet(self,x):
      self.oh = x     #instance variable
   #initializing method
   def __init__(self,x):
      #another instance variable
      self.var = x

#child class def
class MyChild(MyParent):
   #instance method overrides parent method!
```

```
    def mySet(self,x):
        self.oh = x+x   #instance variable

a = MyChild('ok')       #create child instance
#invoke child instance method
#(overriding parent)
a.mySet('hm')
#invoke parent class variable
print(MyChild.info)
#parent instance variable
print(a.var)
#child instance variable
print(a.oh)
```

obj17.py

Here we define `MyParent` with a class variable `info`. We also have an instance method `mySet`, which sets a value for the instance variable `oh`. Finally, we have the `__init__()` method, which sets a value for `var`.

These are all inherited by `MyChild`, but the `mySet()` method of `MyParent` is overridden by a new definition in `MyChild`. We now instantiate `MyChild` with the argument `'ok'`. Since the `__init__()` method is inherited and not overridden, this sets the value of the instance variable `var`. We then invoke the `mySet()` method, but since there is a local version in `MyChild`, that is the version that is used. This means that there is now an instance variable with the value `'hmhm'`. We then print out these values.

Finally, Python allows for multiple inheritance. You can specify more than one class to inherit from. Here's a simple example:

```
class MyMom:            #parent class def
    info = 'wow'        #class variable
    #instance method
    def mySet(self,x):
        self.oh = x   #instance variable

class MyDad:            #other parent class def
    #initializer method
    def __init__(self,x):
        #instance variable
        self.var = x

#child class def multiply inherits
```

```
class MyChild(MyMom,MyDad):
   pass          #nothing in body

#make an instance of the child
#uses initializer from MyDad
a = MyChild('ok')
#instance method from MyMom
a.mySet('hm')
#class variable from MyMom
print(MyChild.info)
#instance variable from MyDad
print(a.var)
#instance variable from MyMom
print(a.oh)
```

obj18.py

Here we define a class `MyMom` with a class variable, an instance method, and a consequent instance variable. We define another class `MyDad` with its own `__init__()` method that creates an instance variable. The class `MyChild` is specified to inherit from both of these, so all of those variables and methods are available to it.

9.4 Syllabification

We illustrate OO in more detail with syllabification. We're going to build up a set of types that we can use to represent syllable structure. We might use these in a model of syllabification, as part of a program that hyphenates, etc.

Let's assume our model of syllabification includes syllables, onsets, nuclei, codas, rhymes, and segments. We might think of a word like *complete* [kʰəm-plit] as being parsed like this:

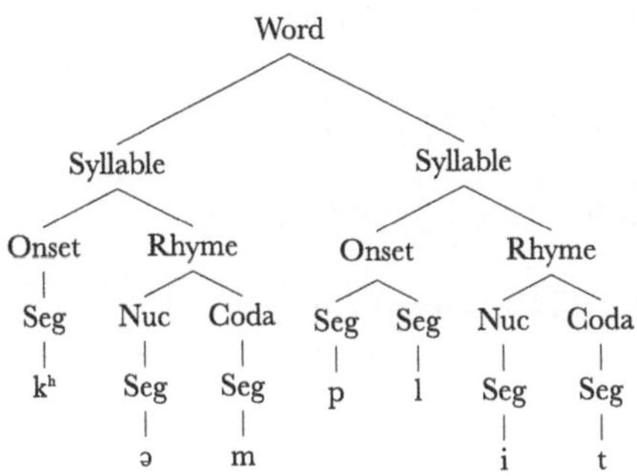

Let's begin with the idea that all of these elements are classes.

```
class Syllable:
   pass
class Onset:
   pass
class Nucleus:
   pass
class Coda:
   pass
class Rhyme:
   pass
class Segment:
   pass
```

obj19.py

It would be reasonable to assume that any element in this hierarchy can be pronounced and spelled. Rather than encode this in each, let's put this into a single class that all of these inherit from:

```
class Speech:           #parent class def
   #initialize with spelling, pronunciation
   def __init__(self,s,p):
      self.spelling = s
      self.pronunciation = p
   def spell(self):  #return spelling
      return self.spelling
   #return pronunciation
   def pronounce(self):
      return self.pronunciation

#all classes inherit from Speech
class Syllable(Speech):
   pass
class Onset(Speech):
   pass
class Nucleus(Speech):
   pass
class Coda(Speech):
   pass
```

```
class Rhyme(Speech):
   pass
class Segment(Speech):
   pass
```

obj20.py

Notice that inheritance here allows for clear savings in that we need not repeat these method definitions for the inheriting daughter classes.

For `Segment`, the lowest level of the hierarchy, this will suffice. We treat `obj20.py` as a module and instantiate it like this:

```
from obj20 import *

a = Segment('ng','ŋ')
print(a.spell())
print(a.pronounce())
```

obj21.py

For higher levels, there are two problems. First, some elements of the hierarchy allow more than one element in a row. For example, an onset can be comprised of zero or more consonants. Another issue is that there is a hierarchy. While onsets are comprised of segments, syllables, etc., they are not directly made up of segments.

Let's address the first issue and revise the arguments to the `__init__()` method. First, we remove it from `Speech`. We then implement it in `Segment` to specify two string arguments.

```
class Speech:          #Speech class def
   def spell(self):    #return spelling
      return self.spelling
   #return pronunciation
   def pronounce(self):
      return self.pronunciation

#everybody inherits from Speech
class Syllable(Speech):
   pass
class Onset(Speech):
   pass
```

```
class Nucleus(Speech):
   pass
class Coda(Speech):
   pass
class Rhyme(Speech):
   pass
class Segment(Speech):
   #initializer from spelling,pronunciation
   def __init__(self,s,p):
      #check that args are strings!
      if type(s) != str or type(p) != str:
         raise Exception(
            'Usage: Segment(str,str)'
         )
      #set instance variables
      self.spelling = s
      self.pronunciation = p
```

obj22.py

We use the `type()` function to check the type for arguments to `__init__()`. We use the `raise` statement to generate an exception, an error, just in case either of the arguments is not a string.

We now turn to the elements that are built on top of segments: onsets, nuclei, and codas. These all have the same general structure in that they are built on segments. Rather than repeat this in each of these classes, we create a new class `Subsyl` that specifies this. The `Subsyl` class will inherit from `Speech` and then `Onset`, `Nucleus`, and `Coda` will inherit from it.

```
class Speech:          #Speech class def
   def spell(self):    #return spelling
      return self.spelling
   #return pronunciation
   def pronounce(self):
      return self.pronunciation

#Segment class def
class Segment(Speech):
   #initialize from spelling and pronunciation
   def __init__(self,s,p):
      #those have to be strings!
```

```
        if type(s) != str or type(p) != str:
            raise Exception(
                'Usage: Segment(str,str)'
            )
        #set instance variables
        self.spelling = s
        self.pronunciation = p

#general class for elements above segments
class Subsyl(Speech):
    #daughter elements are segments
    daughters = Segment
    #initializer from a list of segments
    def __init__(self,xs):
        #set instance variables
        self.spelling = ''
        self.pronunciation = ''
        #got through the argument list 1 by 1
        for x in xs:
            #check that they are segments
            if type(x) != self.daughters:
                raise Exception('Type error!')
            #concatenate spellings
            self.spelling += x.spelling
            #concatenate pronunciations
            self.pronunciation += x.pronunciation

#now inheriting from Subsyl
class Onset(Subsyl):
    pass
class Nucleus(Subsyl):
    pass
class Coda(Subsyl):
    pass
#still inheriting from Speech
class Rhyme(Speech):
    pass
class Syllable(Speech):
    pass
```

obj23.py

Here the `Subsyl` class defines what its daughters must look like: they must be of the class `Segment`. Its `__init__()` method takes a list as an argument. It then checks the type of each element in the list. If they are segments, then it extracts their spelling and pronunciation and adds them to its own spelling and pronunciation.

We can import this and test it out like this:

```
from obj23 import *

c = Segment('c','k')
l = Segment('l','l')
o = Onset([c,l])
print(o.spelling)
print(o.pronunciation)
```

obj24.py

We see that `Onset` functions correctly, inheriting from `Subsyl` directly and indirectly from `Speech`.

Nothing more need be done for `Onset`, `Nucleus`, and `Coda`. We turn then to rhymes and syllables. Rhymes require a nucleus and an optional coda. We capture this with a new `__init__()` method for `Rhyme` that takes an optional argument.

```
class Speech:                #speech class def
   #return spelling and pronunciation
   def spell(self):
      return self.spelling
   def pronounce(self):
      return self.pronunciation

class Segment(Speech):       #segment class def
   #initialize from strings
   def __init__(self,s,p):
      if type(s) != str or type(p) != str:
         raise Exception(
            'Usage: Segment(str,str)'
         )
      self.spelling = s
      self.pronunciation = p

class Subsyl(Speech):        #Subsyl class def
```

```
   #daughters are segments
   daughters = Segment
   #initialize from list of segments
   def __init__(self,xs):
      self.spelling = ''
      self.pronunciation = ''
      #check each element in list
      for x in xs:
         if type(x) != self.daughters:
            raise Exception(
               'Type error for Subsyl!'
            )
         #concatenate instance variables
         self.spelling += x.spelling
         self.pronunciation += x.pronunciation

#inherit from Subsyl
class Onset(Subsyl):
   pass
class Nucleus(Subsyl):
   pass
class Coda(Subsyl):
   pass

class Rhyme(Speech):     #Rhyme class def
   #initialize from Nucleus and optional Coda
   def __init__(self,n,c=''):
      #check type of Nucleus
      if type(n) != Nucleus:
         raise Exception(
            'Type error for Rhyme!'
         )
      #set instance variables
      self.spelling = n.spelling
      self.pronunciation = n.pronunciation
      #check if a coda argument is present
      if c != '':
         #check that it's the right type
         if type(c) != Coda:
            raise Exception(
               'Type error for Rhyme!'
            )
```

```
            #set instance variables
            self.spelling += c.spelling
            self.pronunciation += c.pronunciation

#still inherits from Speech
class Syllable(Speech):
   pass
```

obj25.py

Again, for caution's sake, we test this immediately:

```
from obj25 import *

a = Segment('a','a')
r = Segment('r','r')
t = Segment('t','t')
n = Nucleus([a])
c = Coda([r,t])
r = Rhyme(n,c)
print(r.spelling)
print(r.pronunciation)
```

obj26.py

We create three segments and then build a nucleus and a coda from them. We then assemble them into a rhyme and extract the spelling and pronunciation to make sure everything worked.

We now turn to `Syllable`, which is straightforwardly similar to `Rhyme`:

```
class Speech:              #Speech class def
   #return spelling and pronunciation
   def spell(self):
      return self.spelling
   def pronounce(self):
      return self.pronunciation

class Segment(Speech):  #Segment class def
   #initialize from strings
   def __init__(self,s,p):
      if type(s) != str or type(p) != str:
         raise Exception(
```

```
            'Usage: Segment(str,str)'
         )
      #set instance variables
      self.spelling = s
      self.pronunciation = p

class Subsyl(Speech):   #Subsyl class def
   #daughters are Segment type
   daughters = Segment
   #initialize from list of Segments
   def __init__(self,xs):
      self.spelling = ''
      self.pronunciation = ''
      for x in xs:
         if type(x) != self.daughters:
            raise Exception(
               'Type error for Subsyl!'
            )
         #set instance variables by concatenating
         self.spelling += x.spelling
         self.pronunciation += x.pronunciation

#all inherit from Subsyl
class Onset(Subsyl):
   pass
class Nucleus(Subsyl):
   pass
class Coda(Subsyl):
   pass

class Rhyme(Speech):     #Rhyme class def
   #initialize from Nucleus and optional Coda
   def __init__(self,n,c=''):
      if type(n) != Nucleus:
         raise Exception(
            'Type error for Rhyme!'
         )
      self.spelling = n.spelling
      self.pronunciation = n.pronunciation
      if c != '':
         if type(c) != Coda:
            raise Exception(
```

```
            'Type error for Rhyme!'
         )
      self.spelling += c.spelling
      self.pronunciation += c.pronunciation

class Syllable(Speech):   #Syllable class def
   #initialize from Rhyme and optional Onset
   def __init__(self,r,o=''):
      #check Rhyme type
      if type(r) != Rhyme:
         raise Exception(
            'Type error for Syllable!'
         )
      #set instance variables
      self.spelling = r.spelling
      self.pronunciation = r.pronunciation
      #if onset arg is present
      if o != '':
         #check that it's an onset
         if type(o) != Onset:
            raise Exception(
               'Type error for Syllable!'
            )
         #concatenate with existing
         #instance variables
         self.spelling = o.spelling + \
            self.spelling
         self.pronunciation = o.pronunciation \
            + self.pronunciation
```

obj27.py

The only odd part is that we must order the arguments to `__init__()` for `Syllable` counterintuitively. This is because the onset is optional, and optional arguments must occur on the right. Again, we test this immediately:

```
from obj27 import *

k = Segment('c','kʰ')
a = Segment('a','a')
r = Segment('r','r')
```

```
t = Segment('t','t')
o = Onset([k])
n = Nucleus([a])
c = Coda([r,t])
r = Rhyme(n,c)
s = Syllable(r,o)
print(s.spelling)
print(s.pronunciation)
```

obj28.py

There's obviously a lot more we could do here, but this is sufficient to establish the basics of what object-oriented syllabification might look like.

There's a lot of code here, and it doesn't obviously *do* a whole lot. On the other hand, what it does do, it does clearly. This is typical of an OO approach: it allows you to lay out a higher-level structure and semantics that is clear.

9.5 Exercises

9.1 Rewrite `obj13.py` using functions instead of classes.

9.2 What happens if a method with the same name appears in two classes and some other class inherits from both?

9.3 Give an example showing how multiple inheritance interacts with multi-generation inheritance. Explain how it works.

9.4 In our syllabification example above, the code we wrote for `Syllable` and `Rhyme` is redundant. This could be addressed by enriching the class hierarchy and factoring out the common parts here. Do this.

9.5 Write a class system to handle simple morphology in some language. You will want classes for `Stem`, `Prefix`, `Suffix`, `Morpheme`, and the like.

9.6 Use a class system and inheritance to model historical change. The idea is that languages inherit from other languages but can innovate as well. Build a toy system with the right properties. (This is fun, but tricky. There are lots of ways to make this work.)

9.7 Rewrite the final program from any of the previous chapters using classes. What advantages or disadvantages does this present?

9.8 Write an object-oriented program that parses some amount of HTML. Your program should read in a web page and convert it to a nested set of objects that represent HTML entities. Make sure your program can "print" the web page in some suitable fashion.

9.9 Write an OO program that builds up regular expressions from basic elements. The smallest element should be ϵ and single characters, and your system should allow for union, concatenation, and Kleene star.

9.10 **Web**: Snoop around the web and figure out what the method `__str__()` does. Explain it and write a program that makes use of it.

10 GUIs

So far, all of our programs have been text-based. This means you run them from the terminal window, and the output is either some text in that window or text in some file.

In this chapter we briefly describe how to write Python programs with a *graphical user interface* (GUI). These are programs where your interaction with the program occurs through input modes other than simply typing text, e.g., buttons, menus, dialogs, and other mouse- or trackpad-based operations.

There are many systems you can use to do this with Python, but we will use *tkinter*. There are a number of reasons for this choice.

(i) Tkinter is the oldest GUI for Python and quite stable, so code using it works and is not subject to a lot of version dependencies.
(ii) Tkinter is included in any Python distribution, so it is available regardless of what version of Python you're working with or what your operating system is.
(iii) Tkinter works "out of the box." There are no additional packages required to make it work.
(iv) Tkinter is relatively easy. While it may not have every GUI bell and whistle we might want, it's a good entry point for GUI programming generally.
(v) Finally, there are a number of other languages that have essentially the same system for GUIs, e.g., Perl, Ruby, and Tcl.

There are some limitations of tkinter.

(i) Some versions of tkinter run only if you also have the X11 windowing system installed. (This is available for all operating systems, and you should see a warning if this is the case.)
(ii) Tkinter programs may not have a local "look and feel"; that is, while you can have windows, labels, and buttons, they may not look quite like the normal elements of those sorts on your computer.

(iii) Tkinter is limited. Not all GUI widgets are available.
(iv) Finally, tkinter applications are interpreted programs just like other Python programs, so they will usually not be as fast as other non-Python native GUI applications on your system.

In this chapter we will outline the general logic of GUI programming and how it is done with tkinter. We will cover some of the more usual GUI elements – widgets – and how to enable your program to use those. We conclude with a GUI-version of our stemmer program from Section 7.3.

10.1 The General Logic

We have seen two general models for programming so far. The first is *procedural*. We see programs as a sequence of commands executed in sequence, with control structures to govern that sequence.

The second model, introduced in the preceding chapter, is *object-oriented*. Here we see programs as a network of objects. A program is a set of class definitions, and then, as much as possible, instantiation of those classes makes the program work.

In this chapter, we consider a third model: *event-driven* programming. On this view, we create a set of GUI elements or *widgets*. Those elements are then laid out and "wait" for user input.

The best way to understand this perhaps is to run an example. If you run the following program, it will generate the window below. (This is on a Mac, using X11.)

```
from tkinter import *

r = Tk()          #start tkinter
f = Frame(r)      #make a window
f.pack()
b = Button(       #make a button
    f,
    padx=20,
    pady=10,
    text="Quit",
    command=quit
)
b.pack(           #position the button
    side=LEFT,
    padx=20,
```

```
    pady=20
)
mainloop()      #wait for something to happen
```

gui1.py

The program does nothing until you click/press the button. When you do, the program ends.

The general structure of the code is as follows. First, we import from the `tkinter` module so we can use GUI elements. We then start the GUI with the `Tk()` command. We then create GUI elements: a `Frame` (or window) and a `Button`. We position them with the `pack()` method. In the case of the button, there are a number of variables that control its size, shape, and what it does. In addition, when we invoke `pack()` on it, there are additional variables that govern where it goes. The important point is that one of the variables we specify when we create the button is what it does, in this case `command=quit`.

Once those elements are specified and located on the screen we tell the program to wait for something to happen `mainloop()`. The program simply sits and waits for the user to do something. Any activity of the user is noticed by `mainloop()`, but our program is written so that only a button click in the right location is attended to. In this case, a button click invokes the `quit()` function, which quits the program.

This then is the standard structure for a GUI-based program:

(i) You create a set of GUI elements (widgets).
(ii) The widgets are specified for what functions apply if the user interacts with them.
(iii) You lay your widgets out in some fashion. This includes where they are in the window, whether they are "active" and available for user input, whether they are visible at all, etc.

(iv) Finally, you initiate the event loop, instructing the program to wait for user input.

There are complications that can occur, but this general view will suffice for our purposes.

One other issue is worth mentioning at this point. Tkinter is an interesting system because it actually uses another programming language. The Tcl programming language (Tool Command Language) includes a set of extensions for creating graphical user interfaces called Tk (widget toolkit). Tkinter makes the Tk widgets usable from Python. As we noted above, there are a number of other programming languages that do this as well. The important consequence of this is that GUI commands thus don't always look like other Python commands. Tkinter tries to minimize this, but it's still a fundamental reality of this system.

10.2 Some Simple Examples

In this section we describe a set of basic widgets, their properties, and options for how they can be placed in a window.

The simplest thing you can do with tkinter is nothing. You simply start the GUI system and wait for something to happen. If you do that, you'll get a blank window. The program ends when you close that window.

```
from tkinter import *
Tk()
mainloop()
```

gui2.py

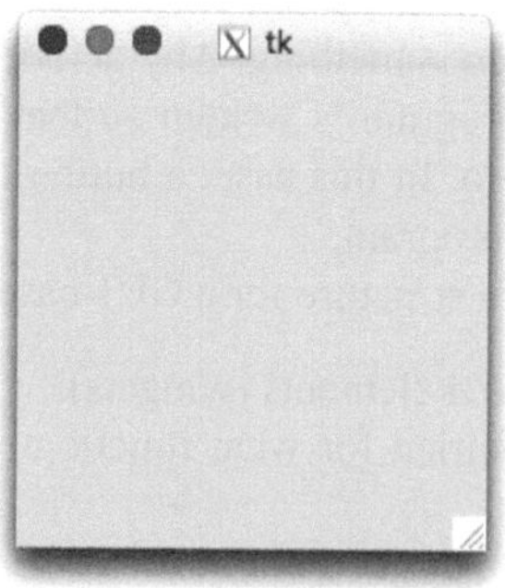

A simple widget that you can add to your window is a label. It allows you to display some text in a window. Here's an example:

```
from tkinter import *

w = Tk()
l = Label(
   w,
   text='Hello world!',
   padx=30,
   pady=30
)
l.pack()

mainloop()
```

gui3.py

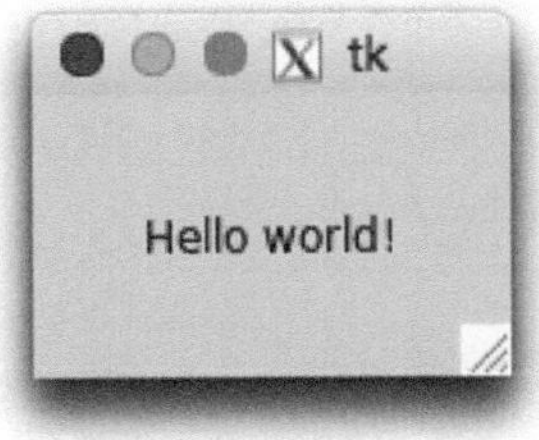

You can also have both labels and buttons:

```
from tkinter import *

w = Tk()
l = Label(
   w,
   text='Press the button!'
)
l.pack()
b = Button(
   w,
   text='Quit',
   command=quit
)
b.pack()

mainloop()
```

gui4.py

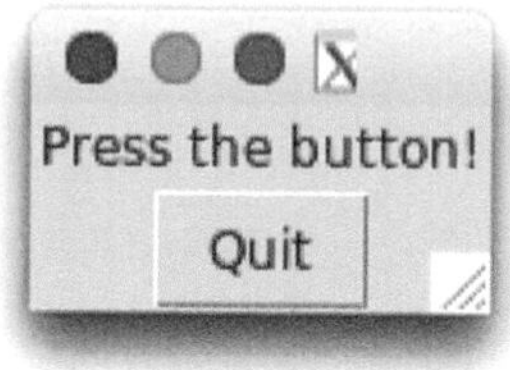

Notice how, in the absence of further specification, buttons and labels are just big enough to hold the text you specify. Similarly, the window is just big enough to hold them. Also note that, unless you specify otherwise, the widgets are laid out top to bottom in the window.

So far, we've seen only buttons that can quit the program, but if you can write a function for it, your button can initiate it. Here's a simple example where pressing a button prints something.

```
from tkinter import *

#function to print something
def wow(): print('wow!')

w = Tk()          #start GUI
l = Label(        #make a label
   w,
   text='Press either button!'
)
l.pack()          #place the label
b1 = Button(      #make a button
   w,
   text='Wow',
   command=wow
)
b1.pack()         #place the button
b2 = Button(      #make another button
   w,
   text='Quit',
   command=quit
)
b2.pack()         #place the other button

mainloop()        #go...
```

gui5.py

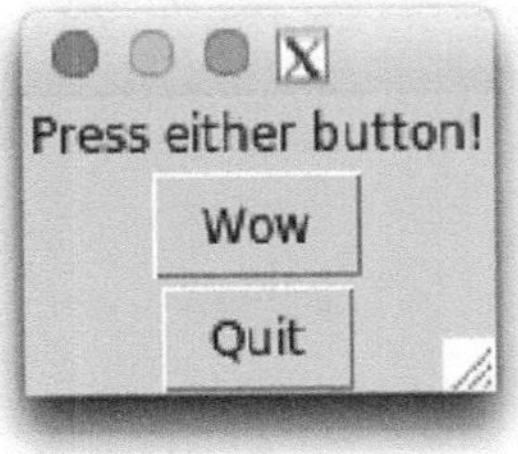

In fact, you can alter the content of your GUI while it's running with a button press by using the `configure()` method:

```
from tkinter import *

#variable to store number of button presses
count = 0

#function to update button presses
def wow():
   #we're updating the variable above, so
   #we have to specify that
   global count
   count += 1     #update the variable
   #reset the label below
   l.configure(text=count)

w = Tk()            #start the GUI
#make and place a label
l = Label(w,text=count)
l.pack()
b1 = Button(        #make and place a button
   w,
   text='Wow',
   command=wow
)
b1.pack()
b2 = Button(        #make/place another button
   w,
   text='Quit',
   command=quit
)
```

```
b2.pack()

mainloop()          #go...
```

gui6.py

Here the upper button changes the value of the label every time you press it. To do this, we first declare a variable `count` and assign it the value 0. We then define a function `wow()` that changes the value of that variable and then assigns the new value as the text of the label. There is an interesting quirk of Python here. If you try to change the value of a variable from outside a function (that you have not passed as an argument to the function) you must declare that it is an external variable with the `global` keyword. Interestingly, even though you are changing the value of the text of the label as well, you need not declare it as global. The rest of the program is as in the previous example.

10.3 Widget Options

Widgets like buttons and labels have a number of shared options that can be set. We won't go through them all here, but they include:

background (bg) The background color.
foreground (fg) The foreground color. In the case of buttons and labels, this is the color of the text.
font What font to display the text in. The font is specified as a triple of name, size, and (optionally) style, e.g., `('times', 14, 'italic')`.

anchor What edge of the widget to align the text to. This only has a visible effect if you force the widget to be bigger than the text. It takes the values: N, NE, E, SE, S, SW, W, NW. The default is CENTER.

command What command to execute, only relevant for buttons so far.

padx How much extra space to leave on the x-axis.

pady How much extra space to leave on the y-axis.

textvariable This can be specified as a particular variable. If the value of the variable changes, the text will change without having to invoke configure().

To see some of these in action, the following program dynamically manipulates foreground, background, and font of a label.

```
from tkinter import *

cs = ['red','blue','green']  #list of colors
fs = [('times',14,'italic'), #list of fonts
('monaco',24),
('Comic Sans MS',30)]

#current foreground, background, and color
fgval = 0
bgval = 0
fontval = 0

#change the *val variables to next leval value
def switch(x):
   if x < 2: x += 1
   else: x = 0
   return x

def fval():                  #change font
   global fontval
   fontval = switch(fontval)
   l.configure(font=fs[fontval])

#change foreground
def fgcol():
   global fgval
   fgval = switch(fgval)
   l.configure(fg=cs[fgval])
```

```
#change background
def bgcol():
   global bgval
   bgval = switch(bgval)
   l.configure(bg=cs[bgval])

w = Tk()                              #start GUI
#make and place a label
l = Label(
   w,
   text='I am a label',
   fg=cs[fgval],
   bg=cs[bgval],
   font=fs[fontval],
   padx=30,
   pady=30
)
l.pack()
#button to change foreground
b1 = Button(
   w,
   text='Foreground',
   command=fgcol
)
b1.pack()
#button to change background
b2 = Button(
   w,
   text='Background',
   command=bgcol
)
b2.pack()
#button to change font
b3 = Button(
   w,
   text='Font',
   command=fval
)
b3.pack()
b4 = Button(
   w,
   text='Quit',
```

```
    command=quit
)
b4.pack()

mainloop()                        #go...
```

gui7.py

First we define a set of three colors and a set of three fonts. (Note that the fonts available may vary from system to system.) We then define three variables to keep track of the current colors for foreground and background and the current font. If you look down at the values for the label, you'll see we use these variables as indices into the lists of colors and fonts. For example, `cs[fgval]` gives the current value of the foreground color. The first three buttons rotate through those values every time you click the button. Thus the three buttons together allow you to see all combinations of these values for foreground, background, and font.

The three buttons call the functions `fval()`, `fgcol()`, and `bgcol()`, which all work the same way, so we'll just talk about `fval()`. When the relevant button is pressed, this function collects the current value of `fontval` which, tells us what the current font is. We then use the `switch()` function to increment the value of `fontval`. If `fontval` is already at its maximum of 2, the `switch()` function sets it to 0. We then use `configure()` to update the font of the label. The other two functions, `fgcol()`, and `bgcol()`, work the same way.

Let's now look at the `textvariable` property. This allows us to dynamically alter the content of a widget without using `configure()`. To use it you must set up a special tkinter variable. When that variable is associated with some widget's `textvariable` and that variable changes, the widget will automatically update without having to invoke `configure()`.

This has to be a special tkinter variable, e.g., an integer `IntVar`, or a string `StringVar`. To use one of these, it must be declared in advance, but after the GUI has started. For example, `x = IntVar()`. In addition, to access the value of one of these you must use a special method, i.e., `x.get()`. Similarly, to set the value, you must use another special method, i.e., `x.set(...)`. The following program exemplifies:

```
from tkinter import *

w = Tk()            #start gui

#create a tkinter integer variable
count = IntVar(w)
count.set(0)        #set it to 0

#function to increment count variable
def wow():
   #get the current value + 1
   c = count.get() + 1
   count.set(c)     #reset count variable

#a label that's tagged for the count variable
l = Label(w,textvariable=count)
l.pack()
b1 = Button(        #a button
   w,
   text='Wow',
   command=wow
)
b1.pack()
b2 = Button(        #another button
   w,
   text='Quit',
   command=quit
)
b2.pack()

mainloop()          #go...
```

gui8.py

Here we declare `count` as an `IntVar` and then set its value with `set()`. The relevant button calls the function `wow()`. That function simply increments the value of `count`, which automatically updates the displayed value of the label.

10.4 Packing Options

We've discussed some simple widgets and their basic attributes. We now turn to laying widgets out in a window or other container.

There are three basic ways to lay things out. The `place()` method allows you to specify precise pixel locations in a window. The `grid()` method splits a window into a grid and lets you place widgets in any specific cell. In this section, we consider the easiest and most common method: `pack()`. This method has a number of basic options:

expand Can be set to `TRUE` or `FALSE`. If `TRUE`, the widget moves to fill available space.

fill Does the widget fill space allocated to it: `NONE` default, `X` fill horizontally, `Y` fill vertically, or `BOTH` fill both.

side What side of the container does the widget pack against: `TOP` default, `BOTTOM`, `LEFT`, or `RIGHT`.

padx, pady How much extra space on the x- or y-axis should there be outside the widget?

ipadx, ipady How much extra space on the x- or y-axis should there be "inside" the widget?

Here is a simple example of a GUI with a label and two buttons packed using default values.

```
from tkinter import *

w = Tk()            #start GUI
#make and place label
```

```
l = Label(text='This is a label')
l.pack()
b1 = Button(       #make and place button
   text='wow',
   command=lambda: print('wow')
)
b1.pack()
#make and place another button
b2 = Button(text='quit',command=quit)
b2.pack()

mainloop()         #go...
```

gui9.py

Note how the three widgets are sized proportionally to the text they display. Notice too that they are organized top-down in the order they were packed. Finally, notice how when we make the window bigger by dragging the lower right corner, the three widgets stay pressed against the upper edge.

Let's now set `expand` to `TRUE` for one of the buttons:

```
from tkinter import *

w = Tk()           #start GUI
#make and place a label
l = Label(text='This is a label')
l.pack()
b1 = Button(       #make a button
   text='wow',
   command=lambda: print('wow')
```

```
)
#place with expand=TRUE
b1.pack(expand=TRUE)
#make and place another button
b2 = Button(text='quit',command=quit)
b2.pack()

mainloop()        #go...
```

gui10.py

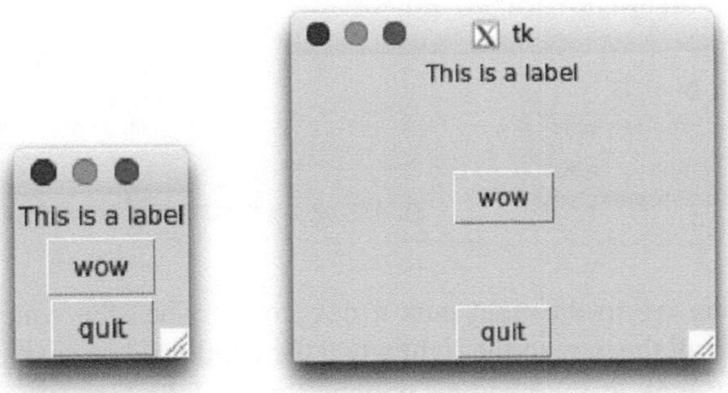

We see no difference when the window initially displays, but when we resize it, we see that the area around `b1` expands to fill the available space pushing `b2` to the bottom.

We can also change what side widgets line up against. Here we set `side` to `RIGHT` for `b1`:

```
from tkinter import *

#all the same except pack for b1
w = Tk()
l = Label(text='This is a label')
l.pack()
b1 = Button(
   text='wow',
   command=lambda: print('wow')
)
b1.pack(side=RIGHT)   #here's the difference
```

```
b2 = Button(text='quit',command=quit)
b2.pack()

mainloop()
```

gui11.py

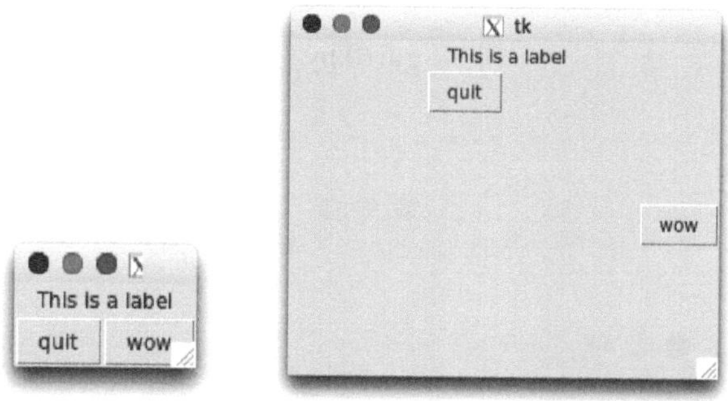

Here, packing `b1` to the right puts it next to `b2` when the window initially displays, but all the way to the right when the window is resized.

Let's now look at the padding options. In the following program, we set `padx` for `b1` and `ipady` for `b2`:

```
from tkinter import *

w = Tk()                #start GUI
#make and place label
l = Label(text='This is a label')
l.pack()
b1 = Button(            #make a button
   text='wow',
   command=lambda: print('wow')
)
b1.pack(padx=30)        #padding on x-axis
#make another button
b2 = Button(text='quit',command=quit)
b2.pack(ipady=20)       #padding on y-axis

mainloop()              #go...
```

gui12.py

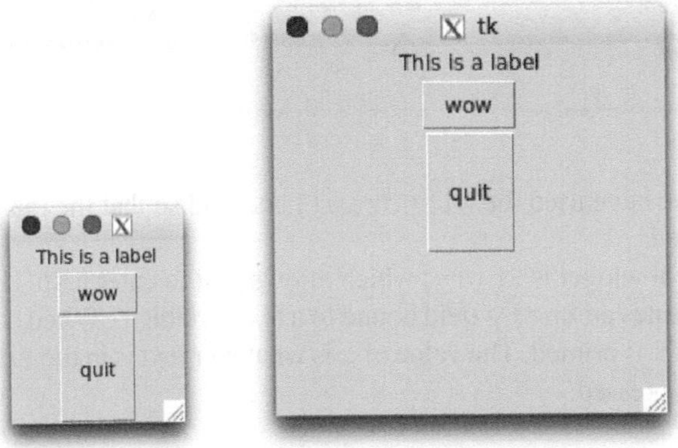

Here `padx` puts extra space to the left and right of `b1`, making the initial window wider. The setting for `ipady` for `b2` makes the button itself taller. When the window is resized, we no longer see the effect of `padx`.

10.5 More Widgets

Let's now consider just a couple more widgets. First, we have the specialized `messagebox` widget. This comes in three varieties: error, warning, and info. The following program exemplifies:

```
from tkinter import *
#special import for messageboxes!
from tkinter import messagebox

root = Tk()                #start GUI
root.withdraw()            #hide main window
messagebox.showerror(      #error message
   "Error",
   "An error occurred"
)
messagebox.showwarning(    #warning message
   "Warning",
   "You are warned!"
)
messagebox.showinfo(       #info message
   "Info",
   "You are informed."
)
quit()                     #quit
```

gui13.py

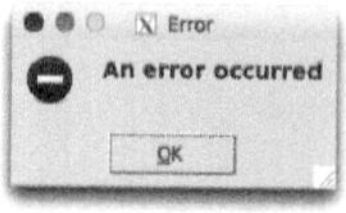

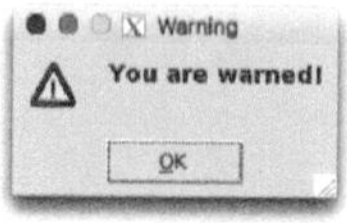

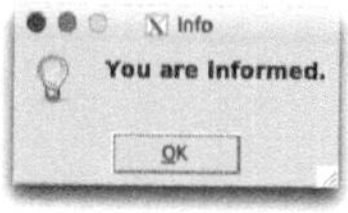

Note that we have used the `withdraw()` method so that the main window is not displayed.

One useful widget is `Entry`, which allows you to enter text. The following program creates an `Entry` field bound to a textvariable `t`. When `b1` is pressed, the value of `t` is printed. The value of `t` is whatever text is in the `Entry` widget when `b1` is pressed.

```
from tkinter import *

#function to print Entry contents
def printit():
   print('Entry:',t.get())

r = Tk()          #start GUI
#set up stringvar AFTER GUI starts
t = StringVar()
#set value of t
t.set('Type something here')
#Entry field linked to t variable
e = Entry(r,textvariable=t)
e.pack()
#button to print contents of Entry
b1 = Button(
   text='Print',
   command=printit
)
b1.pack()
b2 = Button(     #quit button
   text='Quit',
   command=quit
)
b2.pack()

mainloop()
```

gui14.py

One useful feature for an `Entry` widget is to invoke a function if the user types `return` after typing in something. It's a little tricky to do this: you have to *bind* the return to the entry field. One complication is that the name of the return key is `'<return>'`. The other complication is that the `bind()` method automatically passes the "event," in this case `'<return>'`, as an argument to the function. In this case, our `printit()` function does not take an argument, so we have a conflict. To resolve this, we use a `lambda` expression to capture the `'<return>'` argument as `x` and then, effectively, discard it before invoking `printit()`. Here's the code:

```
from tkinter import *

#function to print contents of Entry
def printit():
   print('Entry:',t.get())

r = Tk()           #start GUI
t = StringVar()    #string variable for Entry
#set value of variable
t.set('Type something here')
#Entry field
e = Entry(r,textvariable=t)
e.pack()
#link Entry to return key!
e.bind('<Return>', lambda x: printit())
b1 = Button(       #one button
   text='Print',
   command=printit
)
b1.pack()
b2 = Button(       #another button
   text='Quit',
   command=quit
)
```

```
b2.pack()

mainloop()          #go...
```

gui15.py

The window looks the same here as in the previous case.

10.6 Stemming with a GUI

Recall the stemming program we built in Section 7.3. The final version of the code `manip19.py` starting on page 160 assembled everything together in a function `stem()` and then called that function on a word given as a command-line argument. Here we recast that program in a GUI.

Our first step is to remove the code that runs `stem()` on a command-line argument (the last three lines). This new program, `gui16.py` (not shown here), can then be imported by the GUI we will develop below.

Let's build up our GUI in steps. First, we want a window and a button to quit the program:

```
from tkinter import *
import gui16

r = Tk()
b1 = Button(text='Quit',command=quit)
b1.pack()
mainloop()
```

gui17.py

Let's have a `Label` that displays instructions to the user, an `Entry` that the user will type a word into, another button that will apply `stem()` to the word typed in the entry, and finally another `Label` to display the result:

```
from tkinter import *
import gui16        #import stemmer code

r = Tk()            #start GUI
#label for instructions
linfo = Label(text='instructions...')
linfo.pack()
e = Entry()         #entry for word to stem
e.pack()
#where we'll put the result
lres = Label(text='result...')
lres.pack()
#apply the stemmer code
bstem = Button(text='Stem')
bstem.pack()
#quit
bquit = Button(text='Quit',command=quit)
bquit.pack()
mainloop()          #go...
```

gui18.py

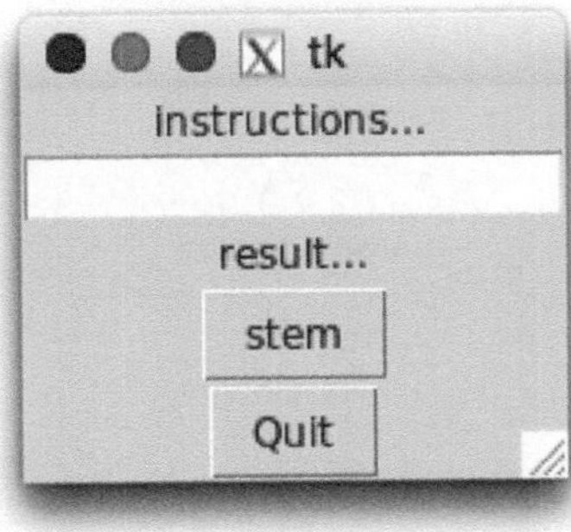

Only the bquit button does anything at this stage, but we can assess whether we have all the elements we need and whether they are laid out to our satisfaction. The following revision tweaks the widgets and their placement to match my own personal esthetic.

```
from tkinter import *
import gui16        #import stemmer code
```

```
f = ('times',18) #a pleasing font

r = Tk()            #start GUI
#instructions in pleasing font
linfo = Label(text='instructions...',font=f)
#put some space on all sides
linfo.pack(pady=10,padx=10)
#entry in the same font
e = Entry(font=f)
e.pack(padx=10)  #a little space on the side
#result goes here in the same font
lres = Label(text='result...',font=f)
#a little space above and below
lres.pack(pady=10)
#the button to do everything
bstem = Button(text='Stem')
bstem.pack()
#quit button
bquit = Button(text='Quit',command=quit)
bquit.pack(pady=10)
mainloop()          #go...
```

gui19.py

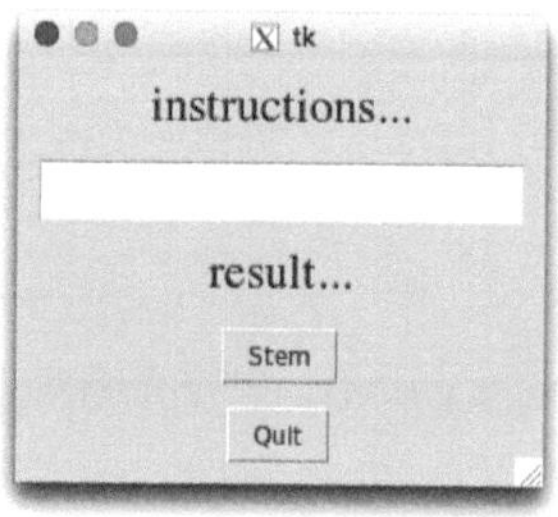

This is a fairly typical step in the GUI programming process. Lay out the GUI to your satisfaction before attaching the real functions you need. A word of caution: it's easy to get distracted by the visual display, making everything look just right in terms of what goes where, fonts, colors, exact placement of widgets, etc.

Let's now add the text of our instructions and textvariables for the `Entry` and second `Label`:

```
from tkinter import *
import gui16     #the stemming code

#more verbose instructions
instructions = '''This is a demo of the
Porter stemmer. Type a
word in the box, press
enter or press the button
and the stem form will be
displayed.'''

f = ('times',18) #that pleasing font again

r = Tk()          #start the GUI

#text variable for the entry
ev = StringVar()
ev.set('Type your word here')
#text variable for the result
lv = StringVar()
lv.set('result....')

#instruction label
linfo = Label(text=instructions,font=f)
linfo.pack(pady=10,padx=10)
#entry variable tied to ev
e = Entry(font=f,textvariable=ev)
e.pack(padx=10)
#result variable tied to lv
lres = Label(textvariable=lv,font=f)
lres.pack(pady=10)
#the button that does everything
bstem = Button(text='Stem')
bstem.pack()
#quit button
bquit = Button(text='Quit',command=quit)
bquit.pack(pady=10)
mainloop()        #go...
```

gui20.py

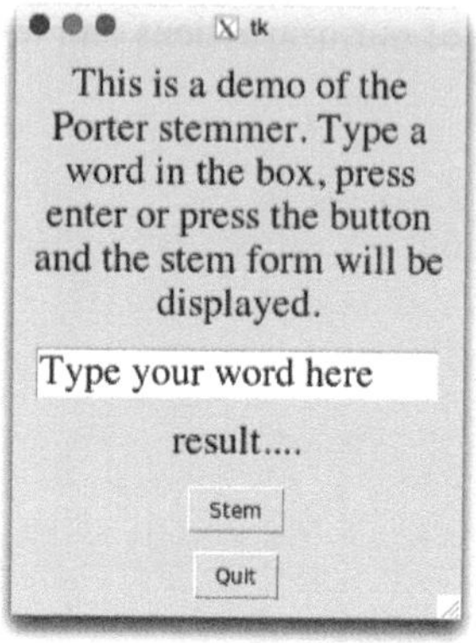

We now add the function called by `bstem` to invoke the stemmer.

```
from tkinter import *
import gui16     #import for stemming code

#instructions
instructions = '''This is a demo of the
Porter stemmer. Type a
word in the box, press
enter or press the button
and the stem form will be
displayed.'''

f = ('times',18) #pleasing font

def guistem():    #invoking stemming function
   w = ev.get()   #get what the user entered
   #stem it
   res = gui16.stem(w)
   lv.set(res)    #display the result

r = Tk()          #start the gui

ev = StringVar() #entry textvariable
ev.set('Type your word here')
lv = StringVar() #result textvariable
lv.set('result....')

#make and place instructoins
```

```
linfo = Label(text=instructions,font=f)
linfo.pack(pady=10,padx=10)
#make and place user entry
e = Entry(font=f,textvariable=ev)
e.pack(padx=10)
#make and place result label
lres = Label(textvariable=lv,font=f)
lres.pack(pady=10)
#stemming button
bstem = Button(text='Stem',command=guistem)
bstem.pack()
#quit button
bquit = Button(text='Quit',command=quit)
bquit.pack(pady=10)
mainloop()        #go...
```

gui21.py

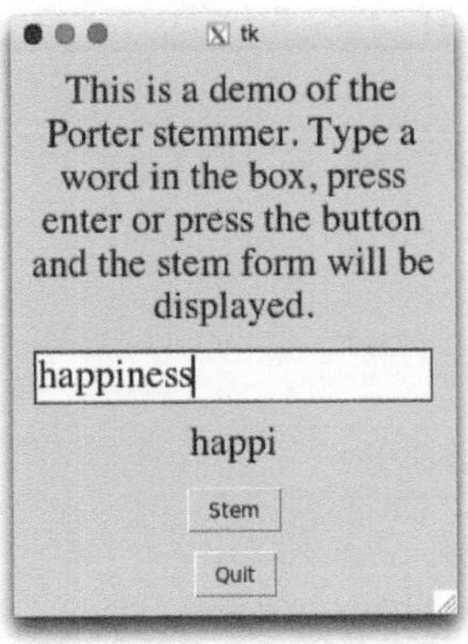

The function is actually quite simple. It collects the content of the `Entry`, applies the `stem()` function to it, and then sets the value of the second label.

Finally, we add some error checking. What happens if the user enters nothing in the entry box, enters more than one word, or enters nonalphabetic letters? We expand the `guistem()` function to respond to these errors.

```
from tkinter import *
#import specifically for messagebox
from tkinter import messagebox
```

```
import gui16     #import for stemming code

#the instructions
instructions = '''This is a demo of the
Porter stemmer. Type a
word in the box, press
enter or press the button
and the stem form will be
displayed.'''

#an error message
error = '''You must enter a single
word with only letters
of the alphabet.'''

f = ('times',18) #the pleasing font

#function for the stemming button
def guistem():
   w = ev.get()  #get what user entered
   #check if it's a single word
   if re.search('^[a-zA-Z]+$',w):
      #if so stem it
      res = gui16.stem(w)
      #display result
      lv.set(res)
   else:             #if not...
      #set result to nothing
      lv.set('')
      #display the error message
      messagebox.showerror('Error',error)

r = Tk()             #start the GUI

ev = StringVar() #textvariable for entry
ev.set('Type your word here')
lv = StringVar() #textvariable for result
lv.set('result....')

#label for instructions
linfo = Label(text=instructions,font=f)
linfo.pack(pady=10,padx=10)
```

```
#entry for the word to stem
e = Entry(font=f,textvariable=ev)
e.pack(padx=10)
#label for result
lres = Label(textvariable=lv,font=f)
lres.pack(pady=10)
#button to trigger stemming
bstem = Button(text='Stem',command=guistem)
bstem.pack()
#quit button
bquit = Button(text='Quit',command=quit)
bquit.pack(pady=10)
mainloop()        #go...
```

gui22.py

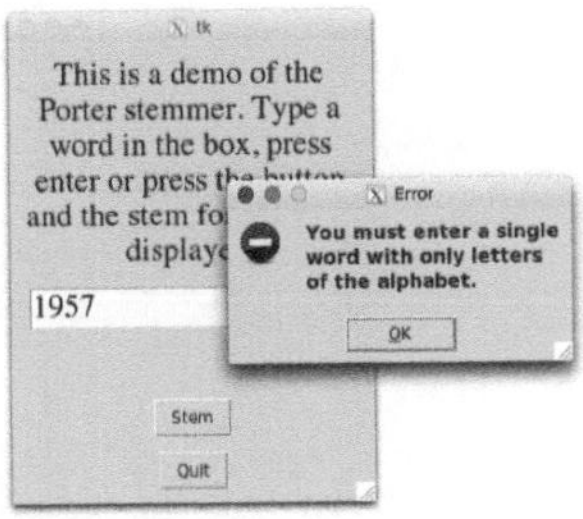

The change is fairly simple. We retrieve whatever is in the `Entry` and check if it is entirely letters of the alphabet. If so, we proceed as before. If not, we set the result `Label` to `''` and we display an error `messagebox`.

10.7 Exercises

10.1 There are a number of other widgets we haven't discussed. Research these on the web, choose one, and write a program that uses it.

10.2 Tweak the code for `gui22.py` so that if the user enters more than one word, each one will be stemmed and displayed.

10.3 Our final program in this chapter created a GUI for a big program from a preceding chapter. Do this for another of those larger programs.

10.4 Buttons can be *active* or *inactive*. Inactive buttons cannot be pressed. Write a program that manipulates this.

10.5 **Web**: There is a datatype called `BooleanVar` that we have not discussed. Find out what it does, explain it, and use it in a program.

10.6 Write a GUI-based program that conducts a psycholinguistic judgment task experiment. Specifically, your experiment will present a set of words to subjects, and subjects will enter the number of syllables they think each word has. Items will be read from a file, and subjects will press buttons to give their response. At the end of the experiment, responses and items are written to a file.

10.7 **Web**: One alternative to the `pack()` method is `grid()`. Find out how it works, explain it, and write a simple program that uses it.

10.8 Write a GUI-based program where there is a single button. When you click the button, it moves to a different part of the window, rotating around the window with each button press.

10.9 **Web**: There are many other GUI systems for Python. Snoop around on the web to find an interesting one, describe how it works, and exemplify it in a program. (Keep it simple here; if you try to create a GUI with a complex system, this can get out of hand quickly.)

11 Functional Programming

In this chapter, we introduce functional programming, a style of programming that grows out of the lambda calculus. Some aspects of functional programming are commonly used in Python, and so it's useful to understand the general principles, if only to make sense of code written by others. This style of programming also lends itself easily to parallel programming and the resulting efficiencies, so this is an additional benefit.

This chapter focuses on advanced material, so if you're new to programming or feel unsure of your foundations at this stage, proceed carefully!

11.1 Functional Programming Generally

Functional programming is an ideal and in practice exists at various approximations to this ideal. The ideal is lambda calculus: *everything* is a function. What this means is two things. First, everything else goes: variables, statements, numbers, strings, etc. Second, functions are really functions; some set of values is mapped to some other value, with no side effects, no variability in what the function maps to.

Python is not the perfect vehicle for functional programming, so the ideal is not completely reached with it, or at least not easily. In practice, what this means is the following:

- We avoid global variables.
- Variables are not used mutably.
- Functions are freely used as arguments to other functions and can be returned from functions.
- Functions can be created anonymously with `lambda` when needed.
- Recursion, overt or covert, is used widely.
- Control structures are avoided or not used at all. Their work is done by functions and comprehensions.

We've covered some of these topics to varying degrees already, but let's go through each of them now in the context of functional programming generally.

11.2 Variables, State, and Mutability

State refers to the value of variables at any particular point in the execution of a program. When we program in pure functional style, the only variables that can exist are those inside functions. When we approximate that style, we reduce the number of variables outside functions and reduce access to those variables. Ideally, any variables outside functions exist only as a convenience, to pass values from each function call to the next. Consider a program like the following:

```
gv = 0              #create global variable

def f1(x):          #function to manipulate global
    global gv
    gv += x

print(gv)           #print initial value of global
f1(3)               #invoke the function
print(gv)           #print the new global value
```

fp1.py

Here, we define a global variable `gv` and assign it an initial value of `0`. We define a function that takes an argument and adds it to the global variable. Since the function manipulates a global variable, we must declare that with the `global` keyword. We then print the initial value of the variable, run the function, and print the updated value of the variable. This produces the following output:

```
> python fp1.py
0
3
```

What this means is that, as we trace the flow of the program, we must keep track of not just what the function might do and the variables it binds, but the state of global variables that exist outside the current function. Contrast this with the following alternative:

```
gv = 0              #create global variable

def f1(g,x):        #function to manipulate global
    g += x
    return g
```

```
print(gv)       #print initial value of global
gv = f1(3,gv) #invoke the function
print(gv)       #print the new global value
```

fp2.py

This produces exactly the same output, but by different means. Specifically, here we do *not* alter the global variable directly in the function. Instead, the function takes the global variable as an argument and then returns the augmented value. As a consequence, the state of the program is local to the function while it is operating. This is a much more functional style and a programming style that is less likely to produce errors. The basic lesson here is not to use the `global` keyword.

Unfortunately, mutability produces essentially the same situation without the red flag of the `global` keyword. Consider the following program:

```
gv = []           #create global list variable
gv.append(0)

def f1(x):        #function to manipulate global
   gv[0] += x

print(gv[0]) #print initial value of global
f1(3)             #invoke the function
print(gv[0]) #print the new global value
```

fp3.py

Here, we've replaced the integer variable with a list. The key difference is that an integer is *not* mutable, but a list *is* mutable. As a consequence, we can refer to the list variable in the function without using the `global` keyword. The effect is exactly the same and the output is the same. As with the first program above, this is to be avoided. Writing code like this means that at any point in our program, we have to inspect what functions are running and what the state of any mutable global variable is to figure out what the program is doing at that time.

As above, we can replace this with code without access to a global (list) variable:

```
gv = []           #create a global list variable
gv.append(0)

def f1(g,x):      #function to manipulate global
```

```
    g[0] += x
    return g

print(gv[0])  #print initial value of global
gv = f1(gv,3) #invoke the function
print(gv[0])  #print the new global value
```

fp4.py

As with `fp2.py` above, we make the global variable an argument to the function and return the changed list variable as the output of the function. The overall output of the program is again the same.

Strictly speaking, we should not have to return the changed value of `gv`; it's a mutable variable, so its value is updated automatically without explicitly returning it. Treating it as immutable like this is a more functional style and less likely to produce errors. The only variables that change value are explicitly changed. In other words, when the `f1()` function *returns* its result, we know precisely what the function has done.

In both cases above, we have moved toward a more functional style by avoiding mutability and manipulating variables globally. Both examples, however, still include a global variable `gv`. We can move to a completely functional style by replacing the global variable with a function as follows:

```
#simulate global variable with a function
def gv():
    return 0

#function that manipulates the global
def f1(g,x):
    return gv() + x

#print initial value of the global
print(gv())
#print the new global value
print(f1(gv(),3))
```

fp5.py

Here we've replaced the global variable `gv` with a function `gv()` that produces the same initial value. We then redefine our `f1()` function to take a function as its first argument and augment its output with its second argument.

Restructuring the program like this means that there are *no* global variables. This may seem like a meaningless technical move, but it has two important consequences. From an abstract perspective, it's a simpler program because there are only functions (and the constant 3). More importantly, from a practical perspective, when we trace the flow of the program from start to finish, we need not keep track of any global variables; we only need to keep track of what the current function is doing. This avoids errors and is much easier to implement using parallel resources, as we'll see in Section 11.8 below.

This final step is not typical of Python programming generally, but it is the ideal in functional programming.

11.3 Functions as First-Class Objects

Another important aspect of functional programming is that functions can be manipulated programmatically. This means you can pass functions as arguments to other functions, and that functions can return functions as values. (We introduced this briefly in Section 5.4.)

The last example above, `fp5.py`, includes a function that can take a function as an argument. The following is a silly example of a function that takes another function as an argument and applies that function to a second argument.

```
x = [5,7,2,6,3]  #arbitrary list of numbers

#a silly function that applies a function
#to an argument
def f1(f,a):
    return f(a)

#testing the function
print('max:',f1(max,x))
print('min:',f1(min,x))
```

fp6.py

Here the function `f1()` simply takes a function as a first argument and applies it to its second argument.

Here's an example of a function that returns a function. Here we are simply converting a string that names a function to that function:

```
x = [5,7,2,6,3]  #arbitrary list of numbers

#a silly function that returns the max
```

```
#function or the min function
def f(n):
   if n == 'max':
      return max
   if n == 'min':
      return min
   return None

#testing the function
print('max:',f('max')(x))
print('min:',f('min')(x))
```

fp7.py

When we allow ourselves to manipulate functions like this, it helps to be able to create functions on the fly. We already introduced `lambda` functions in Section 5.4. To review, we can create an anonymous function with the keyword `lambda` like this: `lambda x: x + 3`. This is an anonymous version of a function that adds `3` to its argument. We invoke it by putting it in parentheses and putting an argument after it in parentheses:

```
>>> (lambda x: x + 3)(2)
5
```

There is, in fact, a set of useful built-in functions that manipulate other functions: `zip()`, `map()`, `filter()`, and `reduce()`. The use of functions like these is typical for functional programming and common in Python.

The `zip()` function takes a set of sequences and constructs a sequence of tuples, drawing one element from each sequence in turn. It returns an object that can be iterated over or that can be converted into a list. Here's an example:

```
x = zip('cat','dog')   #make pairs object
print(list(x))         #convert to list
x = zip('cat','dog')   #make pairs again
for p in x:            #iterate through pairs
   print(p)
```

fp8.py

Here we create a pairs object from the strings `'cat'` and `'dog'`. We convert that object to a list and print it. We then recreate the pairs object and iterate through it, printing each pair. This produces:

```
[('c', 'd'), ('a', 'o'), ('t', 'g')]
('c', 'd')
('a', 'o')
('t', 'g')
```

Note that if we did not create the pairs object a second time, it would be empty when we tried to iterate through it after making a list from it. This is a general property of such objects, and we return to this below in Section 11.7. Note also that if the length of the two sequences is different, the pairing terminates at the end of the shorter sequence. Thus if we tried to zip together `'cat'` and `'dogs'`, we'd get the same output as above.

Another extremely useful function in this family is `map()`. The `map()` function applies some function to every element in a sequence, returning the output for each application in a list. Here's an example:

```
#a list of strings
x = ['Tom','Dick','Harry']
#create a map object where len()
#is applied to each string
y = map(len,x)
#print map object (not too useful)
print(y)
print(list(y)) #convert map object to a list
y = map(len,x) #recreate map object
for i in y:    #iterate through map object
   print(i)
```

fp9.py

Here we create a list of strings and then use `map()` to apply the `len()` function to each element of the list. The output is given below. Notice how printing the result of `map()` directly is not too useful. We must either convert it to a list or iterate through it.

```
<map object at 0x1092744e0>
[3, 4, 5]
3
4
5
```

The `filter()` function is rather similar; it applies a boolean test to every element in a sequence and returns only those elements for which the test returns `True`. Here's an example:

```
#a list of strings
a = ['Tom','Dick','Harry','Mike','Abby']
#create a filter object where a lambda
#function is applied to each element
b = filter(lambda x: len(x) == 4,a)
#convert map object to a list
print(list(b))
```

fp10.py

Here we create a list of strings and then filter it with a `lambda` function. The `lambda` function tests whether the length of a string is four characters long. This returns a `filter` object, which we then convert to a list and print:

```
['Dick', 'Mike', 'Abby']
```

Finally, we have the `reduce()` function from the `functools` module. This function allows you to reduce a sequence of elements to one element by applying a function that takes two arguments. For example, imagine you have a numerical sequence like `[1,2,3,4]` and you apply addition to this. It would first add the first two numbers, producing `3`; then add that to the third number, producing `6`; then add that to the fourth number, producing `10`. Here's a program that does this:

```
from functools import reduce

a = list(range(10))  #create list of numbers
#reduce with addition
b = reduce(lambda x,y: x+y,a)
print(b)               #print result
```

fp11.py

This produces the result `45`.[1]

The `reduce()` function allows for an optional third argument, which is prefixed to the beginning of the sequence before the reduction takes place. For example:

```
from functools import reduce

a = list(range(5))  #create list of numbers
```

[1] The `sum()` function will achieve the same effect directly.

```
#reduce with addition
b = reduce(lambda x,y: x+y,a,9)
print(b)              #print result
#another reudction
c = reduce(lambda x,y: x+y,[],9)
print(c)              #print result
```

fp12.py

The first reduction sums `(((9 + 1) + 2) + 3) + 4`; the second reduction just produces the single default value specified as the third argument: `9`.

11.4 Overt Recursion

A hallmark of functional programming is recursion: functions that are defined with respect to themselves. We've already given an example of this in Section 5.4 above, and we repeat it below:

```
def fac(n):          #function definition
   if n == 1:        #base case of recursion
      return 1
   else:             #recursive clause
      #invokes the function ITSELF
      return (n * fac(n-1))

#invoked with base case
print('1! =',fac(1))
#invoked with recursive case
print('5! =',fac(5))
```

func19.py

Here we define a function to compute factorials. There is a base case for when the argument is `1`. Higher integers are then defined inductively, each one in terms of the next smaller integer. This works because eventually we hit the base case and the self-definition terminates.

Interestingly, most cases of overt recursion like this can be achieved indirectly and more simply using one of the second-order functions we discussed in the preceding section. For example:

```
from functools import reduce

def mult(x,y):          #multiplication function
   return x*y

def fac(n):             #factorial with reduce()
   return reduce(mult,range(1,n+1),1)

#example of the base case
print('1! =',fac(1))
#another example
print('5! =',fac(5))
```

fp13.py

Here we define a function for multiplication and then invoke it with `reduce()` in the definition of `fac()`.

Here's another, more linguistic example. We first define a recursive function for printing out all prefix strings of a word `pfx1()`. The function prints out its current argument and then invokes itself on the same string minus the last character. The function terminates when the string has no characters left.

The nonrecursive version uses `map()`. We define a range of numbers from the length of the string down to `0`. We then apply `map()` to that sequence of numbers to print out all prefixes of the string.

```
#recursive function to find prefixes
def pfx1(s):
   if len(s) > 0:
      print(s)
      pfx1(s[:-1])

def pfx2(s):      #same function using map()
   list(map(
      lambda x: print(s[:x]),
      range(len(s),0,-1)
   ))

pfx1('happy')    #comparing
pfx2('happy')
```

fp14.py

A curious fact about recursion is that, as one might expect, beginning programmers tend to avoid it. Then, as they get more experience, they use it a lot. Interestingly, however, experienced programmers use it less and less, replacing overt recursive techniques with second-order functions like `map()` and `reduce()`.

11.5 Comprehensions

Comprehensions are a powerful tool in Python. They allow you to create lists, sets, or dictionaries from other sequences. They enable us to create various sequences while avoiding overt control structures. The basic syntax for a list comprehension is:

```
[x for y in z if ...]
```

Here, `z` is some sequence of elements, `y` is a variable name, and `x` is a (potentially vacuous) operation applied to `y`. This may be followed by a condition on `y`.

Here's an example:

```
#create list comprehension
a = [x+x for x in 'happy' if x+x != 'aa']
for i in a:       #iterate through list
   print(i)
```

fp15.py

Here we create a list comprehension from the string `'happy'` by concatenating every letter with itself unless that concatenation produces `'aa'`. Here's the output:

```
hh
pp
pp
yy
```

The comprehension can draw from multiple sequences. The following example demonstrates:

```
#create a comprehension from two sequences
a = [x+y for x in 'hat' for y in 'dog']
for i in a:           #iterate through result
   print(i)
```

fp16.py

Here we concatenate two letters where the first is drawn from 'hat' and the second from 'dog'. Here's the output:

```
hd
ho
hg
ad
ao
ag
td
to
tg
```

Notice how the effect is to concatenate every possible combination of letters from the two sequences.

We can nest things in more complex ways as well. The following program exemplifies:

```
a = ['Tom','Ann','Bess']  #list of strings
#nested comprehension
b = [let for wd in a for let in wd]
for i in b:                #iterate through
   print(i)
```

fp17.py

Here we flatten a list of strings into a list of letters by nesting two `for`-clauses. Here's the output:

```
T
o
m
A
n
n
B
e
s
s
```

This structure may seem a bit mysterious, but it makes sense if we translate it into nested `for`-loops:

```
a = ['Tom','Ann','Bess']     #list of strings
b = []                        #nested for loops
for wd in a:
   for let in wd:
      b.append(let)
for i in b:                   #iterate through
   print(i)
```

fp18.py

This produces the same output. Notice how the `for`-loops are nested in the same order as the `for`-clauses in the comprehension.

We can create dictionary comprehensions as well. The syntax is essentially the same except that the entire expression is enclosed in curly braces and the leftmost expression must be a key–value pair. Here's an example:

```
#make a list of strings
a = ['Tom','Dick','Harry']
#dictionary comprehension with strings
#as keys and lengths as values
b = {w:len(w) for w in a}
for w in b:
   print('{:>5}: {}'.format(w,b[w]))
```

fp19.py

Here we first construct a list of strings. We then use that list to create a dictionary comprehension where each string is a key and the value is the length of the string. This produces the following output:

```
  Tom: 3
 Dick: 4
Harry: 5
```

11.6 Vectorized Computation

Second-order functions, recursion, and comprehensions can all be used to avoid control structures and move us toward the functional programming ideal. Another way to do this is vectorized computation or *vectorization*. Computation is vectorized when operations on a sequence of elements can be performed in a single step. Consider the problem of adding together the numbers in two lists. For example:

```
a = [1,2,3]     #create two lists
b = [4,5,6]
print(a+b)      #plus concatenates them
c = []          #make a new list
#go through old lists
for i in range(0,len(a)):
   #add current values together
   c.append(a[i]+b[i])
print(c)        #print result
```

fp20.py

Here we first create two lists of integers. We first see that + concatenates the lists, rather than summing the values. To do that, we must go through the lists and add each pair of elements. The program produces the following output:

```
[1, 2, 3, 4, 5, 6]
[5, 7, 9]
```

One way to avoid this is to use `zip()` and `map()`:

```
a = [1,2,3]             #create two lists
b = [4,5,6]
c = zip(a,b)            #zip each pair together
#sum elements of each pair
d = list(map(sum,c))
print(d)                #print result
```

fp21.py

This produces the same result.

The `numpy` module provides this same functionality for all mathematical operations. What we do is convert our lists to `numpy` arrays, and then mathematical operators have their vectorized effects. For example:

```
import numpy as np

#create two numpy arrays
a = np.array([1,2,3])
b = np.array([4,5,6])
```

```
c = a + b          #vectorized addition
d = a * b          #vectorized multiplication
print(c)           #print results
print(d)
```

fp22.py

Here we cast our lists as `numpy` arrays. We then add and multiply them together and print the results:

```
[5 7 9]
[ 4 10 18]
```

This is quite convenient and efficient but works for only mathematical operations. For string operations, we must use either the usual control structures or solutions like the one above with `zip()` and `map()`.

11.7 Iterables, Iterators, and Generators

In this section, we look more closely at iteration. We start with *iterables*. These are data structures that can be iterated over with control structures like `for` or `while` and include lists, dictionary keys, strings, and files. All of these can be used with higher-order functions as well. For example:

```
from functools import reduce

a = [1,2,3,4,5]                #lists
#add pairs
print(reduce(lambda x,y: x+y,a,0))

#dictionaries
b = {'Tom':14,'Dick':25,'Harry':8}
#length of each key
print(list(map(len,b)))

c = 'happy'                    #strings
#double each letter
print(list(map(lambda x: x+x,c)))

f = open('test.txt','r')   #files
#length of each line
```

```
print(list(map(len,f)))
f.close()
```

fp23.py

An *iterator* is a special kind of iterable. They come in two forms. One form uses OO programming and is based on the `__next__()` method:

```
#an iterator that represents a sequence
#of integers
class Sequence:
   #initialize endpoints
   def __init__(self,low,high):
      self.now = low
      self.high = high
   #required for iterators
   def __iter__(self):
      return self
   #gets the next item
   def __next__(self):
      #must throw this error at limit
      if self.now > self.high:
         raise StopIteration
      #return current value and increment
      else:
         i = self.now
         self.now += 1
         return i

s = Sequence(3,10) #create a new sequence
#iterate through the sequence
for i in s:
   print(i)
```

fp24.py

Here we define an iterator as a new class `Sequence`. To work as an iterator, it must have several key bits. First, it must have an `__iter__()` method that returns the class instance. Second, it must have a `__next__()` method that returns the next element in the sequence. Third, if the sequence is bounded, the `__next__()` method must `raise` a `StopIteration` error when the bound is reached.

The other kind of iterator is built around the `__getitem__()` method. The basic idea here is to build a list that we can index into. Here's an example:

```
#a different kind of iterator
class Students:
   #initialize with list of names
   def __init__(self,listofstudents):
      self.students = listofstudents
   #how to change a name
   def __setitem__(self,n,name):
      self.students[n] = name
   #how to retrieve names
   def __getitem__(self,n):
      return self.students[n]
   #returns number of names
   def __len__(self):
      return len(self.students)
   #adds a name to the list
   def append(self,name):
      self.students.append(name)
   #deletes an item
   def __delitem__(self,n):
      del self.students[n]

#create a new student list
s = Students(['Mike','Joey'])
s[1] = 'Diane'     #change name of 2nd student
s.append('Bob')    #add a student
del s[0]           #delete a student
for n in s:        #iterate through final list
   print(n)
```

fp25.py

The essential method here is `__getitem__()`, but we've added a few others for extra functionality. The `__setitem__()` method lets us change an item. The `__len__()` method lets an instance of the class respond appropriately to the `len()` function. We define `append()` so that we can append names on the end of the list and `__delitem__()` so that we can delete items.

There are virtues in both approaches. The `__getitem__()` approach allows us to use indexing with our iterator, which would seem to be an advantage. On the other hand, the `__next__()` approach allows us to have *infinite* iterators. Here's a simple example:

```
#an iterator that represents an
#infinite sequence of integers
class Sequence:
   #initialize starting point
   def __init__(self,low):
      self.now = low
   #required for iterators
   def __iter__(self):
      return self
   #gets the next item
   def __next__(self):
      #return current value and increment
      i = self.now
      self.now += 1
      return i

s = Sequence(3)      #create a new sequence
for i in range(5): #iterate through sequence
   print(next(s))
```

fp26.py

Here we tweak the code from `fp24.py` by eliminating the endpoint. We can now iterate through a `Sequence` object infinitely. Note that we do *not* want to convert one of these to a list with `list()`. Since the list would be infinite, this does not terminate!

Here's an interesting example where we use the same technique to calculate the Fibonacci numbers. This is the sequence of numbers where the first two numbers are 1 and 1, but each subsequent number in the series is the sum of the preceding two numbers, i.e., $\langle 1, 1, 2, 3, 5, 8, \ldots \rangle$.

```
class Fibonacci:   #an iterator for Fibonacci
   #initialize starting point
   def __init__(self):
      self.now = [1,1]
   #required for iterators
   def __iter__(self):
      return self
```

```
    #gets the next item
    def __next__(self):
       #return current value and increment
       i = self.now[-2]
       self.now.append(
          self.now[-2]+self.now[-1]
       )
       return i

f = Fibonacci()    #create a new sequence
#iterate through sequence
for i in range(10):
   print(next(f))
```

fp27.py

Here we populate the internal list with two numbers. The trick is that we return the second number from the right on every instance of `__next__()`.

Iterators are useful but, as you can see, take a fair amount of work to code. There is another method available, though: *generators*. Generators typically involve less coding but can be mysterious. Compare the following with `fp24.py` above:

```
def sequence(low,high): #sequence generator
   while low <= high:
      yield low
      low += 1

s = sequence(3,10)          #create new sequence
for i in s:                 #go through sequence
   print(i)
```

fp28.py

Generators are functions with two differences. First, the `return` keyword is replaced with `yield`. Second, if the generator function is called a second time, execution resumes *where it left off the first time*. In the case above, this means that the `while` loop inside the `sequence()` function continues right after the `yield` statement, thus incrementing the value of `low` and starting the loop again.

If we think of `yield` as analogous to `return`, then a generator function looks rather odd with statements *following* the `yield` statement. This makes sense, however, if we keep in mind that, on a subsequent application of the function, execution resumes *after* the yield.

The same technique will work to generate the Fibonacci series:

```
def fibonacci():   #fibonacci generator
   res = [1,1]
   while True:
      res.append(res[-1]+res[-2])
      yield res[-3]

j = 0                 #iterate through values
for i in fibonacci():
   print(i)
   j += 1
   if j == 10: break
```

fp29.py

Here we build the sequence just as we did with fp27.py above. The difference is that we embed this in an infinite while loop in the generator function. Every time the function is invoked, the loop resumes. Again, notice that the generator code is much more succinct than the OO-based iterator technique.

11.8 Parallel Programming

Most modern personal computers have multiple processors or cores. What this means is that they can do more than one thing at a time. Python by default does not take advantage of multiple processors, but the multiprocessing module lets you do this. This can result in dramatic improvements in how long your programs run.

Not all tasks lend themselves to parallel programming. Even tasks that can be recast in parallel must be carefully structured to do this. It turns out that adopting a functional programming style makes this relatively easy.

Imagine you have a programming task that you are interested in using parallel programming for. The key question is whether you can break it into subtasks that can proceed independently of each other. This is not always an easy thing to determine.

Consider, for example, reading in a file and building a concordance of the words in that file. It would be difficult to parallelize the file-reading part of this task. The problem is that, no matter how many processors your computer has, it probably has only one hard drive. If you tried to parallelize the file-reading part of the program, you'd have bottleneck from the one hard drive.[2]

Let's assume then that the file has been read in with a single process. Is there any way to parallelize the process of making a concordance? If we had a single process, we would define a dictionary and would then go through the words of the file one by one, adding words and/or incrementing counts for each word as we went. Viewed like this, this would be a difficult task to parallelize as well. The problem here is that the dictionary is a bottleneck. Each process would need to be able to make changes to a single shared dictionary. We would need to make sure that the independent processes did not create conflicts. For example, imagine two separate processes attempted to augment the value for some entry at the same time.

To solve challenges like these, we need to separate our program into separate tasks that are not subject to hardware or software bottlenecks. Accessing the same shared drive at the same time is a hardware bottleneck. Accessing the same dictionary at the same time is a software bottleneck.

In the hypothetical example we've been discussing, the way to proceed might be as follows:

(i) Read in the file.
(ii) Break it into words.
(iii) Partition those words into *n* separate lists.
(iv) Construct separate concordances as dictionaries for each list *in parallel.*
(v) Merge the dictionaries.

Notice how we have solved both the hardware and software bottlenecks with this plan. The hardware bottleneck is resolved by reading in the file in a single process. The software bottleneck is resolved by having separate dictionaries for each of the parallel processes.

While the strategy above resolves the bottlenecks, it's not actually the best way to proceed. There are two efficiency issues. The first is how much time we would save by this process. Breaking our dictionary construction process into separate smaller processes that can proceed independently will surely save time. On the other hand, there is overhead. First, we have to break our list of words into separate lists, which takes time. Second, once we have constructed these separate dictionaries, we have to merge them together, which also takes time. It may well be the case that these overhead steps cancel out the advantages of the parallel steps.

There is a second, cruder, efficiency question to address as well. Imagine that indeed parallelizing like this saves us 10 minutes in runtime. Imagine, however, that this program will need to be run only once and that it takes you half an hour

2 Solid-state hard drives do not have this property and we exploit this below.

to convert your program into one that will run in parallel. This is probably not a good tradeoff.

As a general rule of thumb then, you can parallelize some process when you can efficiently break it into subtasks that can proceed simultaneously without hardware or software bottlenecks and that can efficiently be merged together again.

From a practical perspective, you can parallelize your program if some or all of it can be treated with `map()`. Recall that `map()` allows you to apply some function to some sequence. If you can write your code with `map()`, then you can parallelize your program by replacing `map()` with the method of the same name from the `multiprocessing` module.

We've already given an example of this in Section 8.4 with a program that retrieves webpages. The program is repeated below:

```
import time       #for timing info
#to read webpages
from urllib.request import urlopen
#to do more than one thing at once
from multiprocessing import Pool

#current time in milliseconds
def mytime():
   return round(time.time() * 1000)

def myget(url):  #50 characters of a webpage
   start = mytime()
   data = urlopen(url,timeout=5).read()[:50]
   result = {"url": url, "data": data}
   now = str(mytime() - start)
   print(url + ": " + now + "ms")
   return result

#some random urls
urls = ['http://www.google.com/',
'http://www.yahoo.com/',
'http://golwg360.cymru/newyddion',
'https://news.google.com/news',
'https://tartarus.org/martin/PorterStemmer/',
'https://en.wikipedia.org/wiki/Main_Page',
'http://www.u.arizona.edu']

#print urls in order accessed
```

```
for i in range(len(urls)):
   print(i+1,': ',urls[i],sep='')
print()

mypool = Pool()  #multiple processes
start = mytime() #start the clock
#separate process for each url
results = mypool.map(myget, urls)
#print total elapsed
now = str(mytime() - start)
print("Total = " + now + " ms\n")
```

web11.py

This program retrieves several web pages in parallel using `map()` from the `multiprocessing` module. The key here is that each retrieval can proceed independently. There is no software bottleneck because the different retrievals do not share any data structures. There is no hardware bottleneck either because each retrieval depends on *other* computers. In fact, given that the other computers can take unpredictable amounts of time to respond, the task lends itself quite well to a parallel implementation.

This same technique would seem to work with our hypothetical concordance example above. Here is a nonparallel version of it.

```
import re,time

def mytime():     #gets time in milliseconds
   return round(time.time() * 1000)

start = mytime() #start the clock

#read file
f = open('alice.txt','r')
t = f.read()
f.close()

t = t[11000:]     #strip header

t = t.lower()     #normalize
t = re.sub('[^a-z]+',' ',t)

#split into words
```

```
ws = t.split(' ')

d = {}              #build concordance
for w in ws:
   if w in d:
      d[w] += 1
   else:
      d[w] = 1

print(len(d))
print(sum(d.values()))

#print total elapsed
now = str(mytime() - start)
print("Total = " + now + " ms")
```

fp30.py

This program retrieves the contents of the *Alice* text, strips the header info, normalizes it, breaks it into words, and then uses a dictionary to build a concordance. We've added code to time the execution. On my own laptop, this takes about 15 msec.

Here now is a parallelized version of the same code:

```
import re,time
import multiprocessing as mp

def mytime():       #gets time in milliseconds
   return round(time.time() * 1000)

start = mytime()    #start the clock

def myconc(wds):    #build concordance
   d = {}
   for w in wds:
      if w in d:
         d[w] += 1
      else:
         d[w] = 1
   return d

#read file
```

```
f = open('alice.txt','r')
t = f.read()
f.close()

t = t[11000:]        #strip header

t = t.lower()        #normalize
t = re.sub('[^a-z]+',' ',t)

ws = t.split(' ') #split into words

#partition words
ws = [ws[:9000],ws[9000:18000],ws[18000:]]

#start multiprocessing
mypool = mp.Pool()
res = mypool.map(myconc, ws)

d = res[0]           #merge results
for dx in res[1:]:
   for w in dx:
      if w in d:
         d[w] += dx[w]
      else:
         d[w] = dx[w]

print(len(d))
print(sum(d.values()))

#print total elapsed
now = str(mytime() - start)
print("Total = " + now + " ms")
```

fp31.py

The difference here is that after we have split the text into words, we partition those words into three lists and then use `map()` from `multiprocessing` to make three separate dictionaries *in parallel* from those lists. We then assemble those dictionaries into a single dictionary. On my own laptop, this takes about 45 msec.

As we speculated above, any advantage in parallelizing is offset by other costs here, either the time taken to split the list into three sublists, the time taken to merge the dictionaries, or the time taken by Python to manage the multiprocessing. The moral is that parallel processing does *not* always help.

If we tilt the balance, then the parallel treatment becomes the better one. Following is an example of a program that searches through a set of program files looking for words where pairs of letters are repeated in overlapping fashion. For example, *prepend* repeats the letters *e* and *p* in this way.

```
import re,time,glob

def mytime():      #time in milliseconds
   return round(time.time() * 1000)

start = mytime()  #start the clock

#names of files that end in .py
filenames = glob.glob('*.py')

def countwords(filename):
   f = open(filename,'r')
   t = f.read()
   f.close()
   t = t.lower()
   t = re.sub('[^a-z]+',' ',t)
   ws = t.split()
   wds = []
   for w in ws:
      m = re.search('(.).*(.).*\\1.*\\2',w)
      if m:
         wds.append(w)
   return len(wds)

#map function to all filenames
wds = map(countwords,filenames)
print(sum(wds))    #print total
#stop the clock
now = str(mytime() - start)
#total elapsed time
print("Total time = " + now + " ms")
```

fp32.py

First, we use the `glob()` function from the `glob` module to list all files matching a specific pattern. We then define a function `countwords()` that reads a file, normalizes the text, splits it into words, and finds the words that fit the interlocking pattern described above. This pattern matching requires back-references and is rather time-consuming. We then use the `map()` function to apply the `countwords()` function to each filename and collect the results.

Following is a parallel version of the same program. Here, we simply replace `map()` with the method of the same name from the `multiprocessing` module.

```
import multiprocessing as mp
import re,time,glob

def mytime():        #time in milliseconds
    return round(time.time() * 1000)

start = mytime()   #start the clock

#names of files that end in .py
filenames = glob.glob('*.py')

def countwords(filename):
    f = open(filename,'r')
    t = f.read()
    f.close()
    t = t.lower()
    t = re.sub('[^a-z]+',' ',t)
    ws = t.split()
    wds = []
    for w in ws:
        m = re.search('(.).*(.).*\\1.*\\2',w)
        if m:
            wds.append(w)
    return len(wds)

#start multiprocessing
mypool = mp.Pool(processes=4)
#map function to all filenames
wds = mypool.map(countwords,filenames)

print(sum(wds))    #print total
#stop the clock
```

```
now = str(mytime() - start)
#total elapsed time
print("Total time = " + now + " ms")
```

fp33.py

Here, running these on the program files from the text, the parallel version of the program performs slightly better on my own laptop. In principle, we might expect a hardware bottleneck because of the hard disk access in the `countwords()` function, but my hard disk is solid state, which, in principle, allows parallel access.

11.9 Making Nonsense Items Again

Let's now use some of the techniques we've learned here to tackle the problem of generating nonsense items again as we did in Chapter 3. Let's try something a little more difficult though and build a function that will give us an *infinite* number of nonsense items.

The idea would be to have a function that takes a list of letters as an argument and then generates every possible combination of those letters with no upper bound.

To make an infinite sequence, we need a generator, so the basic template for our function would be like this:

```
def items(ls):
    ...
    yield item
    ...
```

We define a generator function `items()` that takes a string argument `ls`. The function would then yield individual items on each application.

We'll do this recursively. We define a base set of strings `['']` and then generate every concatenation of this with the letters in the function argument. We yield those one at a time. When we've exhausted that list of strings, we redefine that list as the base, concatenate every member of the new base with the function argument, and loop.

For example, if our argument was `'ab'`, then every concatenation with the base gives `['a','b']`. We yield each of those and then do the concatenation again giving `['aa','ab','ba','bb']`. To do this concatenation we use a list comprehension.

This is exemplified in the following program. The `items()` function is as described. We then iterate through the first ten items just as we did above when we introduced generators. We also make use of the `islice()` function from the `itertools` module, which lets us extract a sequence of values from the generator.

```
import itertools

#generate all combinations of letters
def items(ls):
   n = ['']
   while True:
      nn = [x+y for x in n for y in ls]
      for x in nn:
         yield x
      n = nn

j = 0   #iterate up to a specific point
for i in items('abc'):
   print(i)
   j += 1
   if j == 10:
      break

#extract a sequence from the iteration
myslice = itertools.islice(items('ab'),100,105)
print(list(myslice))
```

fp34.py

11.10 Exercises

11.1 Convert a program from an earlier chapter to a fully functional version of the same program. There should be no variables outside of functions (unless they are immediately passed to the next function).

11.2 Function composition of two functions `f` and `g` is when we create a new function that does the same thing as applying `f` to the output of `g`. Write a function that does this. Your function should take two arguments: the two functions. Your function should return a new composed function.

11.3 Write a program that calculates the sum of the number of characters in some set of text files. Your program should take the name of a file directory as a name. Your program should use the `reduce()` function from the `functools` module to do this.

11.4 Take a program from the chapter on control structures that has nested loops and rewrite it using a list comprehension.

11.5 Root Mean Square (RMS) Amplitude is the square root of the average of the squared values of a waveform. Write a function that loads in a wave and calculates this using vectorized calculations from the `numpy` module.

11.6 Write a generator function to create nonsense words. The function should take a list of syllables as input and return any number of concatenations.

11.7 Find a programming problem that is more efficiently solved with parallel programming. Demonstrate that this is so with a sequential and a parallel version of the same program.

11.8 Create an iterator class for syllables in English. Your class should be able to be instantiated with the IPA representation of a word and you should be able to index into it for the syllables of the word. Note that this entails that, in the background, your class will syllabify the word when the class is instantiated.

11.9 Create an iterator class for phrases. You should be able to instantiate the class with a string to be tokenized into words. You should create additional specific classes for NP, VP, etc. that inherit from your `Phrase` class and that can be assembled into a `Sentence`.

11.10 **Web**: There is a function `iter()` that can be useful for functional programming. Snoop around on the web, find out what it does, explain, and exemplify in a program.

Appendix A
NLTK

The Natural Language Toolkit (NLTK) is a huge, publicly available module that includes a number of extremely useful tools for working with language. It also includes a number of text resources that you can use to try out these tools or for your own research.

In this appendix, we very briefly introduce NLTK. Our goal is to give a brief overview of what you can do with it. To find out more, you should consult the module website: `http://nltk.org`.

A.1 Installing

Installing NLTK is straightforward with `pip` or `conda`. The trick is that after you do the install, you should install additional optional components. To do this, go to the interactive environment and type:

```
import nltk
nltk.download()
```

This will open a separate window that lets you choose additional components to install. Installing all the optional components takes slightly less than two gigabytes (version 3.2.3).

A.2 Corpora

Once you've installed the optional components, a number of corpora are installed as well. You have access to them through the `nltk.corpus` submodule. For example, the `names` corpus is a corpus of 7,944 English personal names separated into male and female categories. The following program shows how to load these and then calculate the average number of letters in male and female names:

```
from nltk.corpus import names
```

```
male = names.words('male.txt')
female = names.words('female.txt')

mavg = sum(map(len,male))/len(male)
favg = sum(map(len,female))/len(female)

s = '{:>6} names: {:>5} {:>3}'
print(s.format(
    'Male',
    len(male),
    round(mavg,3)
))
print(s.format(
    'Female',
    len(female),
    round(favg,3)
))
```

nltk1.py

First, we import the corpus from `nltk.corpus`. We then extract the male and female names into separate lists, calculate averages for each, and print the results.

Many other corpora are available as well. If you want to see what's available, go to the interactive environment and type:

```
from nltk.corpus import names
names.root
```

This will import the `names` corpus and then report where it is located on your system. Go to that directory and scroll through the directories listed under `corpora`. Some of the files here are helper programs for `nltk.corpus`, but most are corpora. Here are a few that I have on my own system:

- `inaugural` Fifty-six presidential inaugural addresses
- `gutenberg` Eighteen novels, plays, and poetry collections from Project Gutenberg
- `brown` A tagged version of the Brown corpus: 1,161,192 words
- `shakespeare` Eight of Shakespeare's plays
- `timit` A sample of the TIMIT corpus including recordings and transcriptions of 160 sentences
- `wordnet` A dictionary resource indicating a variety of semantic relationships between words for English

There are many others.

A.3 Tokenizing

One of the simplest things you can do with a corpus is to tokenize it, break it into words or sentences. This has already been done for many of the corpora that come with NLTK. The following code loads the `brown` corpus and prints out the first five sentences and the first five words:

```
from nltk.corpus import brown

words = brown.words()  #get the words
for w in words[:5]:    #print first 5 words
   print(w)
#get the sentences
sentences = brown.sents()
#print the first 5 sentences
#(with extra spaces)
for s in sentences[:5]:
   print(s,'\n')
```

nltk2.py

NLTK also offers utilities for tokenizing your own texts. The following code loads the *Alice* text and extracts the first 10 sentences:

```
from nltk.tokenize import sent_tokenize

f = open('alice.txt','r')  #read in alice
t = f.read()
f.close()
t = t[11000:]              #strip header
#break into sentences
ss = sent_tokenize(t)
#how many sentences
print(len(ss))
#print first 10 with extra spaces
for s in ss[:10]:
   print(s,'\n')
```

nltk3.py

You can also tokenize words. The following code normalizes and tokenizes the words of the *Alice* text and then plots the most frequent ones:

```
import nltk
from nltk.tokenize import word_tokenize

f = open('alice.txt','r') #read in alice
t = f.read()
f.close()
t = t[11000:]                #strip header
t = t.lower()                #normalize
ws = word_tokenize(t)        #break into words
#create a frequency distribution
fd = nltk.FreqDist(ws)
#print the 50 most frequent items
fd.plot(50,cumulative=False)
```

nltk4.py

Here we use the `word_tokenize()` function to extract the words and the `FreqDist` class to create the distribution. We use the `plot()` method of that class to plot the 50 most frequent items. (Run the code yourself and notice the zipfian distribution.)

The `word_tokenize()` function can take an optional second argument, which allows tokenization in a number of other languages as well.

A.4 Stop Words

The frequency display for `nltk4.py` above is dominated by function words. We can remove these from the display using the `stopwords` corpus. This includes sets of function words from a number of languages. For example:

```
import nltk
from nltk.corpus import stopwords
from nltk.tokenize import word_tokenize

f = open('alice.txt','r') #read in alice
t = f.read()
f.close()
t = t[11000:]                #strip header
t = t.lower()                #normalize
ws = word_tokenize(t)        #break into words
#get stop words
```

```
sws = stopwords.words('english')
#remove stop words
ws = [w for w in ws if not w in sws]
#create a frequency distribution
fd = nltk.FreqDist(ws)
#print the 50 most frequent items
fd.plot(50,cumulative=False)
```

nltk5.py

Notice that this does not remove punctuation.

We can use this to do fairly crude language identification. The basic idea is to check a bit of text to see how much it overlaps with the different stopword sets. The following program exemplifies:

```
from nltk.corpus import stopwords
from nltk.tokenize import wordpunct_tokenize

#sample sentences
s1 = 'Here is a short sentence for you.'
s2 = 'Voici une phrase courte pour vous.'
s3 = 'Вот короткое предложение для вас.'

def identify(s):     #to count stop words
   print(s)
   wds = wordpunct_tokenize(s.lower())
   wdsset = set(wds)
   for l in stopwords.fileids():
      stopwordsset = set(stopwords.words(l))
      score = len(
         wdsset.intersection(stopwordsset)
      )
      print('\t',score,l,
         len(stopwords.words(l))
      )

identify(s1)        #test on the 3 sentences
identify(s2)
identify(s3)
```

nltk6.py

A.5 Tagging

Finally, NLTK offers several mechanisms to obtain or produce tagged text: text with part of speech indicated.

The simplest is to use one of the corpora that are already tagged, e.g., the `brown` corpus. The following code shows how to get tagged sentences from the `brown` corpus:

```
from nltk.corpus import brown

for s in brown.tagged_sents()[:5]:
   print(s,'\n')
```

nltk7.py

Here each sentence is output as a list of pairs where the first member of the pair is a word and the second member is a part of speech tag.

NLTK can also tag text for you. The following code shows this for the first few sentences of *Alice*:

```
from nltk import pos_tag
from nltk.tokenize import sent_tokenize
from nltk.tokenize import word_tokenize

f = open('alice.txt','r')  #read in alice
t = f.read()
f.close()
t = t[11000:]              #strip header
#break into sentences
ss = sent_tokenize(t)
#go through the first 5 sentences
for s in ss[:5]:
   ws = word_tokenize(s)    #break into words
   print(pos_tag(ws),'\n') #tag and print
```

nltk8.py

A.6 Summary

In this appendix we've presented some easy things you can do with the NLTK module. This only scratches the surface.

Index

Printed in the United Kingdom by TJ Clays Ltd.